Reading Borough Council
34126002332065

Mismatch

mismatch

Why our World No Longer Fits our Bodies

PETER GLUCKMAN
AND
MARK HANSON

OXFORD
UNIVERSITY PRESS

Great Clarendon Street, Oxford OX2 6DP

Oxford University Press is a department of the University of Oxford.
It furthers the University's objective of excellence in research, scholarship,
and education by publishing worldwide in

Oxford New York

Auckland Cape Town Dar es Salaam Hong Kong Karachi
Kuala Lumpur Madrid Melbourne Mexico City Nairobi
New Delhi Shanghai Taipei Toronto

With offices in

Argentina Austria Brazil Chile Czech Republic France Greece
Guatemala Hungary Italy Japan Poland Portugal Singapore
South Korea Switzerland Thailand Turkey Ukraine Vietnam

Published in the United States
by Oxford University Press Inc., New York

First published 2006

British Library Cataloguing in Publication Data
Data available

Library of Congress Cataloging in Publication Data

Gluckman, Peter D.
Mismatch : Why our bodies no longer fit our world
Peter Gluckman and Mark Hanson.
p. cm.
Includes bibliographical references and index.
ISBN-13: 978-0-19-280683-3 (alk. paper)
ISBN-10: 0-19-280683-1 (alk. paper)
1. Ecological genetics. 2. Evolution (Biology). 3. Genotype—environment
interaction. I. Hanson, Mark. II. Title.
QH456.G58 2006
576.5′8—dc22 2006019107

Typeset by RefineCatch Limited, Bungay, Suffolk
Printed in Great Britain
on acid-free paper by
Clays Ltd., St Ives plc

ISBN 0–19–280683–1 978–0–19–280683–3

1

Foreword

At the end of the first chapter of *The Life and Opinions of Tristram Shandy,* Mrs Shandy asks 'Pray, my dear, have you not forgot to wind up the clock?' to which Tristram's distracted father responds: 'Good G—! Did ever woman, since the creation of the world, interrupt a man with such a silly question?' And once he is born, Tristram is trapped thereafter in a time-warp, his whole life (and the book about him) muddled by time. As he says, 'I wish either my father or my mother, or indeed both of them, as they were in duty both equally bound to it, had minded what they were about when they begot me; had they duly considered how much depended upon what they were then doing.'

Lawrence Sterne was writing in 1759. But this idea—that events at the time of conception, or during gestation, influence not only the outcome of a pregnancy but also the developing character of the child—has prevailed since earliest written records. There is evidence for this in cuneiform tablets from Sumer and in papyri from ancient Egypt. The Biblical account in Genesis when Jacob encourages the birth of speckled sheep (probably a genetically recessive characteristic) in Laban's flock of pure whites, by showing them black and white whittled sticks at the moment of conception, suggests that this notion has been widely prevalent for a very long time indeed.

The discoveries of Gregor Mendel, the gardening monk from Brno who died in 1884, led to a more 'rational' understanding of inheritance. But the implications of his work were not fully accepted until well after his death, more than forty years after his original observations. His detailed experiments with around 28,000 pea plants eventually gave birth to the idea of the gene being the unit of inheritance, with the Laws of Inheritance named after him. That insight, and the rising importance of the theory of evolution promulgated by Charles Darwin, who died just two years before Mendel (and who may have just possibly known of his controversial work), led to a radical change in our ideas of inherited characteristics.

These ideas were not universally accepted very quickly. It took until the

1920's when it became widely agreed that the genetic variation in populations occurred because of mutations, or changes, in genes. Phenotypic changes, i.e. changes in the constitution or appearance of an organism, were seen to be gradual as a response to the pressures of natural selection which selected those inherited characteristics that better matched the organism with its environment. Subsequently the modern science of molecular biology showed how the structure of the DNA with its base-pair sequences influences the phenotype by producing specific changes in the RNA and the manufacture of various proteins. And it also became clearer how the maternal and paternal chromosomes, carrying the genes were passed to their offspring.

But this knowledge, once accepted, led to rather determinist notions about human genetics. It was very easy to think of a gene consistently producing just one aspect of the phenotype, but this is far from the complete story. Genes can produce their influence in ways that are much more subtle than was widely understood. So it is only quite recently that the conventional, deterministic view of genetic inheritance has had to be re-evaluated. In the last fifteen years or so there has been increasing evidence that the environment can have a much greater effect on the way genes work than was realized in Mendel's original concept. There is startling evidence, too, that the environment that prevails during an individual's early development can radically affect later life. These early influences may be particularly important when the individual is still inside the uterus. There is strong evidence that many of these effects may be caused by chemical changes in genes and may immediately alter the characteristics of succeeding generations.

One compelling, recent discovery was that disease in middle or old age could have its origins in events before birth. David Barker, from the University of Southampton, meticulously trawled through a huge number of records of the births of children delivered in Hertfordshire just before, during, and after the First World War. Most of the mothers of these children had normal pregnancies. But inevitably, there were some hospital records of women with complicated pregnancies. Some of their babies experienced a deficient environment while in the uterus and were not well nourished, being born well below normal birth weight in consequence. In general, the babies who were born much smaller than average particularly those who were really small, around two kilograms in weight—turned out to be at increased risk of ill health in later life. A far greater proportion of the adults who started life as very small babies died from coronary heart disease before the age of 65 years. The beginning of their life had been spent in a sub-optimal milieu and the effort of attempting to compensate for this left them with constitutional scars which had grave effects decades later. These babies were, in effect, mismatched to their early environment and had tried to adapt to survive. David Barker and his colleagues subsequently showed that heart disease was not the only risk. They were also more likely to suffer the related diseases of stroke, high blood pressure, and diabetes by the time they reached middle age.

Even more remarkable are observations that the environment during early development can produce striking effects in later generations. In 2001, Lars Bygren and his colleagues in Sweden published studies showing how some boys born in 1905 seemed adversely affected by their grandparents' diet. In particular, their grandfathers' access to copious food, or lack of it, affected the longevity of these boys. In an isolated, remote part of northern Sweden there had been regular crop failures and bumper harvests in different years between 1799 and 1880, and these had been very well documented in public records. If the paternal grandfather had had access to copious food from a bumper crop during the time when he went through puberty, his grandchild —if male—was more likely to die at a younger than average age. No cause and effect has been firmly established. Yet the higher incidence of diabetes in these grandsons argues that a plentiful diet produced some chemical changes in the genes on the Y chromosome of their grandparents' germ cells during a critical stage of early development. These changes could have affected how the genes in their sperm expressed when they produced their children. So the environment of the grandparent had a deleterious effect on the boys (who carried that Y chromosome) two generations later.

This book is in part about these intriguing influences arising during development, which so far have received too little attention. It places these within the context of how humans evolved, because we now live in a very different world from that which our species first inhabited. The book breaks new ground in our understanding and will be essential reading for anybody interested in the fascinating complexities of human biology.

It is a privilege to write the foreword to this book by Peter Gluckman and Mark Hanson, two eminent researchers who have managed to describe their ideas in an accessible and entertaining manner. The book is a most timely and innovative contribution to the popular debate about genes and the environment. It is now becoming increasingly clear that how individuals are matched or mismatched with their early environment and how they adapt can have the most profound effects on their health. Those effects may not be apparent until much later in life, and may affect the health of future generations. Though we now smile at Tristram Shandy's curious predicament, perhaps the ancient intuitive understanding of what influences developing humans was not so wide of the mark. The knowledge and research described in this valuable book presents compelling scientific evidence about this strange but crucial aspect of our evolution and development.

Robert Winston

April 2006

Preface

This book started thirty-five years ago, when as young trainees we went on separate medical expeditions, one to Africa, the other to the foothills of the Himalayas. Through those visits we learnt that significant health problems can arise in a population through an unfortunate combination of susceptibility to disease coupled with an environment which increases the risk of that disease. What we saw on those two expeditions directed us both to careers in medical research as a means to improving human health.

Independently, we focused our careers on a particular aspect of biology: development. This led us on other journeys of discovery, from research in classical physiology, through developmental biology, and now to evolutionary biology and attaining a new perspective on gene–environment interactions. It taught us how the processes of evolution and development come together to allow humans to live well in some situations and not so well in others. We felt the need to synthesize our ideas for a wider audience, and this book is the result.

This book is also about journeys. They include the journey that the human species has undertaken in its evolution, but also the journey that every one of us has taken through our development from conception. When we travel across time zones by air, it takes several days for our biological clocks to readjust. Similarly if groups of people go on journeys by migrating to new environments, their evolutionary and developmental biology may not be able to cope well with the change and perhaps will not be able to readjust for several generations. How well people adapt to such transitions primarily depends on whether their biology can meet the variety of challenges that their environment poses, and this reflects how well matched they are to that environment. The greater the degree of mismatch, as the authors saw so clearly in people living in Nigeria and Nepal all those years ago, then the greater the risk of disease. But many problems which humans now face arise from less dramatic examples of mismatch. Understanding this concept can lead to new approaches to old problems, and so we believe it to be very important.

So we decided to travel again, now between our medical research laboratories in Auckland and Southampton, to write a book about match and mismatch—their origins and their consequences.

Contents

Introduction: Our Bodies and our World *1*

Part I. Match

1. Our Comfort Zone *17*
2. Where Have We Come From? *49*
3. When We Were Very Young *74*
4. Things Ain't What They Used to Be *94*
5. Constrained by our Pasts *124*

Part II. Mismatch

6. Coming of Age *137*
7. A Life of Luxury *158*
8. Four Score Years and Ten *178*
9. Match and Mismatch *194*

Epilogue *212*
Notes *213*
Acknowledgements *267*
Index *269*

Introduction
Our Bodies and our World

'How on earth can anyone bloody-well live here? All I want to do is collapse and die!'—this is almost inevitably your first thought if you climb or trek to an altitude above 3,500 metres. Unless you are well prepared by having taken the time to become acclimatized, all you can do is take two or three steps, then pause for some gasping breaths before taking another two or three steps, then more gasping and another two or three steps . . . it never seems to end. You are totally unsuited—*mismatched*—to being at such a high altitude. Yet the reality is that people do live at such high altitude, and even higher, in the Himalayas and in the Andes, and have done so for many generations.

The Sherpa are one such people. They are of Tibetan origin but crossed the Himalayas to live in the high valleys of Nepal hundreds of years ago. Until they were 'discovered' by the great Himalayan climbing expeditions of Mallory, Hunt, and others in the middle of the twentieth century, they lived in almost total isolation. Essentially their only contacts came through trading expeditions, particularly to obtain rock salt, into Tibet. The Sherpa are Buddhist subsistence farmers living on potatoes, barley, and yaks.[1] In more recent years porterage to support climbing expeditions and trekking tourism has changed their economy, but at a considerable cost to their traditional society and environment. The influx of tourists has caused deforestation because the slowly growing mountain trees have been used for firewood. For some reason the tourists do not like their food cooked the traditional Sherpa way over dried yak dung—although frankly the food really tastes much the same however it is cooked over a smoky fire. The point is that the physical and social environment of the Sherpa has changed rapidly.

But in 1972, tourism had not yet arrived in the upper Khumbu valley which leads up to Mount Everest and down which the Dukh Khosi river tumbles past the famous Thangboche monastery and Namche Bazaar, the biggest Sherpa village. Thirty years ago, this valley was really only known to the climbing community and to a few people working with Sir Edmund Hillary, the New Zealander who with Sherpa Tensing Norgay was the first to conquer Everest in 1953. Hillary then chose to devote much of his life to supporting the Sherpa people through the development of schools and basic infrastructure, including a small hospital, bridges, an airstrip, and engineering works to protect some of the monasteries. One of us (Peter) travelled there as a newly graduated doctor to assist a medical research expedition initiated by Hillary. The aim of the expedition was to study major health problems resulting from iodine deficiency—these were the price the Sherpa unwittingly paid for living in this extreme environment with its steep terrain and high snowfall. While these lofty valleys might be free from the inter-tribal conflict and inadequate pastures which had led their ancestors to migrate over the mountains several hundred years earlier, the environment posed other major problems. We will tell the story of this expedition because it illustrates some important points about human ability to adapt to the environment, and because it sets the scene for *Mismatch*.

Over millions of years the Himalayan mountain range, which is very young in geological terms, has been formed by the collision of the Indian and Asian tectonic plates. The Indian plate pushed under the Asian plate as it moved inexorably northwards, driven by convection in the Earth's mantle, generating earthquakes and pushing up the mountains. As the Himalayas grew they were subject to repeated deluges of rain and snow. This washed the soil almost completely free of some minerals, in particular iodine. The result is that the Himalayan foothills are perhaps the most iodine-deficient region of the world. All humans require some iodine in their diet, but this was not possible for the Sherpa in their mountain environment. Even in the mid-twentieth century, they were almost totally isolated from access to western foods supplemented with iodine.

The expedition that Peter joined found that over 90 per cent of the Sherpa population had goitre. This is the medical term for a grossly enlarged thyroid—this is a gland located in the neck just below the voice-box which manufactures and secretes a hormone called thyroxine (or thyroid hormone) into the bloodstream. This hormone is made from a combination of an amino acid and iodine, the iodine being absorbed from the diet by the gut. Thyroxine is essential for normal function because it determines the body's rate of metabolism—acting in some ways like the accelerator of a car. If the accelerator is pushed down too far (by having too much thyroxine) the engine (the body's metabolism) revs too fast. If there is not enough thyroxine, the body's metabolism slows down and does not have enough power to function properly. So the result of the low iodine in the diet was that many Sherpa had slow metabolism. The goitres that Peter saw in the Sherpa were sometimes larger than the neck itself, giving a grotesque appearance. Unfortunately enlarging the gland does not solve the problem—even a bigger gland cannot make thyroxine without adequate iodine in the diet. And so the inevitable signs of slowed metabolism appeared, such as delayed reflexes, fluid retention, higher blood fat levels, and poor heat generation by the body.

The secretion of thyroid hormone is controlled by a very elegant system involving another hormone, thyrotropin, made by the pituitary gland, which is the so-called 'master gland' at the base of the brain. If the thyroid hormone levels in the blood are low, more thyrotropin is secreted and the thyroid gland is driven to secrete more thyroid hormone. Conversely if thyroid hormone levels are high in the blood, less thyrotropin is secreted. This kind of control system is called a negative-feedback loop and is a common way of maintaining constancy within a biological system.[2] It is analogous to a thermostat-controlled heater—as a room gets colder, the thermostat signals to the heater that it must generate heat; when the temperature has risen to the preset level the thermostat turns off, but it clicks on again if the room starts to cool. This is not an entirely closed system because you can change the setting on the thermostat, and in just the same way biological systems acting through the brain's control of the pituitary gland can alter the requirements for thyroid hormone. But if thyroid hormone levels remain low for a long time, for example because the diet does not contain sufficient iodine, the continuously high levels of thyrotropin also stimulate growth of the thyroid gland in a desperate attempt to make more thyroxine. As the gland enlarges, it becomes visible in the neck as a goitre.

Of even greater concern to the doctors on the expedition was that one in eight of the population showed a particularly tragic consequence of iodine deficiency, one that started before birth. In some people the lack of iodine during fetal life drastically affected their brain development. They were born as cretins. This is a medical term (unkindly used in a pejorative way by people who have not witnessed the condition) for the severe mental retardation associated with intrauterine iodine deficiency. Yet these *Kurs* (the

Sherpa word for these individuals) were generally well integrated into society and had valuable jobs. They were the water carriers—spending their days carrying buckets of water from streams in the valleys to houses on the high terraces.

Despite much that was known about cretinism at the time of the expedition, there was still a major scientific mystery. Not all Sherpa, despite their very low iodine intakes, showed these signs of iodine deficiency. And even more striking was the fact that not all babies were born cretins even though virtually all their mothers were iodine deficient. On closer examination the mystery deepened still further. Some cretins had a particular form of cerebral palsy which resulted from their iodine deficiency interfering with brain development in an irreversible way. But others did not.[3] Some cretins were extremely dwarfed (less than 1.4 metres tall when fully grown) but others were not. And some were deaf-mute whilst others were not. All these various clinical pictures appeared to be different manifestations of the same environmental deficiency: lack of iodine. And when Peter and his fellow doctors treated potential mothers in this population with iodine injections, all the different forms of cretinism disappeared, showing the central role of iodine deficiency in producing them.

Lessons from the mountains

Here were some important lessons for a young medical scientist. The first was that not all individuals show the same symptoms and signs, even when faced with the same conditions. Clearly the potential for goitre, thyroid hormone deficiency, and cretinism was the result of an interaction between some intrinsic susceptibility, perhaps based on individual variations in genetic make-up, and the environment in which they lived. The source of such variations was not always clear. It might be the individuals who varied, or their environment, or both. For example one village, Phortse, had a particularly high rate of cretinism. Phortse was located away from the other villages and made a beautiful picture after the snow had fallen, with its two-storeyed stone houses—yaks living below, people living above—scattered across ancient stone terraces on the side of the valley. Was the particular problem in Phortse due to some local genetic variation predominant in the families that lived there? Or was it due to their diets being slightly different from the other villages? On investigation the only difference seemed to be that the villagers of Phortse ate much more barley. While the doctors could not prove it, they thought that the barley might contain something that interfered with thyroid gland function. This was not implausible because in Tasmania children had been shown to develop goitre in the spring, but it then disappeared later in the year, only to return the following year. The effect turned out to be due to contamination by a chemical in cow's milk which came from the wild turnip that appeared in the pasture every spring.[4]

So the messages were clear—the environment of a population can vary in subtle ways which may not be immediately obvious, but which nonetheless can have dramatic effects on the pattern of disease in that population. Not everyone has the same constitution and so not everyone responds to any particular environment in the same way. The more we look, the more we realize that this principle applies not just in Nepal, but across the world. And even subtle changes in the environment can have a major impact, depending on the nature of the change and when in the person's life course it occurs. This variation in our constitution and our ability to match our biology to our environment is central to understanding how we live in this world, and whether we remain healthy or develop disease.

The Sherpa were paying a severe price in terms of their health for living in the high Himalayan valleys. But in many other ways as a population they had adapted astonishingly well to this rugged place, and that is why they stayed there. They had overcome the problems of breathlessness (unlike Peter!) and could carry enormous loads to high altitudes. While the normal porter's load was 30 kg, they were able to carry a double load—and be paid twice as much. One of the Sherpa porters on the expedition would regularly carry a load of over 60 kg—a good deal more than he weighed himself. He had clearly developed strong muscles as well as lungs.

But it was not just in terms of their anatomy and physiology that the Sherpa had adapted to their environment. They had also developed over many generations a complex social structure and a sophisticated culture. This culture differs in many ways from that to which most of us are accustomed. For example, some practised polyandry, the practice of a wife having several husbands. Even more surprisingly, often her multiple husbands were brothers. Such customs appear to assist the Sherpa in coping with their extreme environment, in this case because the wife needed to have a strong man around at home even if her other husbands were off tending the yak herd.[5]

For Peter's expedition, the urgent question was what could be done to deal with the problem of iodine deficiency in the Sherpa. This was not easy to answer. In Europe the solution would be to add iodine to a foodstuff, as had been done with the introduction of iodized powdered table salt. Such measures, taken in the 1920s, quickly prevented goitre in parts of the UK. The condition of 'Derbyshire neck' was no longer seen. The 'fashionable' distended throats of young women who sat as models of doomed heroines for the Pre-Raphaelite painters of the middle of the nineteenth century became a thing of the past. But while powdered salt might have seemed the appropriate way to bring iodine to the Sherpa, the researchers found that Sherpa tradition demanded that they use rock salt brought over the high altitude passes (about 5,800 metres) from strife-torn Tibet. There were several strong cultural reasons which could be identified for continuing such trade: it preserved communication between groups of the same religion; and it was part of the expectation of the young males that they should face the dangerous passage across the icy storm-weathered high altitude passes. Such perspectives had to be understood in attempting to find a solution to the problem. The doctors ended up by injecting each Sherpa with a depot of iodine, a treatment costing only about 10 cents, which lasts at least five years. It is a solution that has been adopted in several other remote mountainous areas such as New Guinea where food-based alternatives such as iodized salt cannot be reliably applied.

When we think about such lessons, drawn from working and living with the Sherpa in the Himalayas, some more general biological insights appear, and they form a starting point for this book. The first is about ability to adapt (adaptedness). Clearly humans can live in some very extreme environments, well away from the savannah of central Africa where we first evolved. We can become matched to a variety of environments because, like some other animals such as the rat and cockroach, we have broad adaptive capacity. In contrast many other species of animal are exquisitely matched to a particular environmental niche. No one can doubt how well the emperor penguin is adapted for a life fishing in frigid ocean waters and breeding on the Antarctic ice shelf; or how well the cheetah is suited for sprinting at over 100 km/hour to run down an impala on the savannah; or indeed how well the chameleon or the stick insect have perfected the art of camouflage to make them well hidden from predators in their habitat.

The second point relates to cost. The polar bear is not adapted for life in the tropics nor the Malayan sun bear for life in the Arctic. If they were to be transported from one to the other, they would perish. Similarly, humans have evolved to be able to live in a broad range of environments, but nonetheless we are not infinitely adaptable. Although we can often cope at least for a while when we get to the extremes of this range, if we try to live beyond these environmental limits there will be a cost. So a species thrives if it lives in an environment for which its 'design' matches it. The greater the degree of mismatch between environment and design, the greater the cost. In this book we call this the *mismatch paradigm*, and the cost of such mismatch is often disease, just as the cost to the Sherpa of inhabiting the high Himalayan valleys was goitre and cretinism. As we carried out our research, we wondered whether even the so-called 'modern environment' in which so many of us aspire to live lies beyond the limits of human adaptedness and, if so, how? Could there be a cost even for contemporary societies living in developed countries? When we realized the implications of the answer to this question, we decided to write this book.

Design for a life[6]

'Design' is a term used extensively in developmental and evolutionary biology but unfortunately it has been compromised by its adoption by the Intelligent Design, anti-evolutionist movement and, as a result, has been somewhat misused.[7] Our body's 'design' is shaped by both the genes we inherit and the processes of our development from an embryo to an adult. Both the processes of evolution, which selected the genes that make us what we are, and the processes of development are dependent on interactions between the organism and its environment. They operate over very different time-scales—evolution usually over thousands of years,[8] development over less than one lifetime. So understanding design requires an understanding of how genes, environment, and development interact.

Nor is there a need to get bogged down in teleology.[9] When we use words like 'design' and 'choice' or 'strategy' in this book we do not intend them to have any implications of conscious intent. They are simply code words, a kind of shorthand to explain how evolution and development work. Design is the net outcome of the interaction of evolutionary influences and the processes of development, which results in a mature organism with its particular characteristics. 'Strategy' refers to the pathway an organism may take in attempting to match its biology to its environment, and 'choice' to situations where the actual pathway chosen has been influenced by other factors such as environmental cues. These are no more conscious choices than is your 'choice' to breathe harder if you undertake exercise or go up to high altitude in the Himalayas.

Making the match

Humans have chosen to live in an enormous range of environments. The Sherpa live at high altitude, the Inuit above the Arctic circle, the Fuegians—described in rather uncomplimentary terms by Darwin in his book *The Voyage of the Beagle*[10]—lived at an extreme southerly latitude, the Tuareg live in the Sahara desert . . . these are all examples of human existence in extreme natural environments. Other populations live in threatening environments created by humans themselves. Parts of Southern California and the Australian outback are becoming more difficult to live in as the underground aquifers are drained to provide water for cities and industry and the increasingly salty soil cannot support plant growth. Nauru Island in the Pacific sustained a stable society for over a thousand years until colonial powers removed all the topsoil to mine the nitrate-rich guano in their voracious desire for fertilizer—from being a rich vibrant society on a luscious tropical island, Nauru is now a horrific landscape of bedrock and there is discussion about relocating the remaining islanders to Australia or New Zealand. The Easter Islanders lived (until they nearly all died out) with the consequences of tree destruction on their remote Pacific island—they could no longer build boats and so fish could not be part of their diet.[11] Other Polynesians had to live with the consequences of overcrowding on small islands and some populations went as far as widespread infanticide to control their numbers. The Japanese fishermen of Minamata Bay had to live with the consequences of mercury poisoning which caused tragic brain damage to their children.[12] Some soldiers who served in Vietnam may live with their exposure to Agent Orange and it is possible that these effects are transmitted to their offspring. The list of tragedies goes on and on. They are all examples of how, even without migrating to remote places, we really can make our environment challenging to inhabit.

The most serious consequences were seen in the Sherpa when the iodine deficiency occurred during fetal life, as this led to cretinism. Similarly in

Minamata Bay and the Vietnam veterans it was the impact of the environmental crisis on the developing fetus or infant which had the most dramatic consequences. This is a major theme of this book—that the environment we are exposed to during our early development from a single fertilized egg, to an embryo, fetus, and infant, can have long-term consequences.

The developmental perspective has had surprisingly little influence in forming our understanding of the human condition. Yet embryology was a very important component of the nineteenth-century biologists' research, including that of Darwin himself.[13] He recognized that the complexities of embryonic development might reveal much about how different species evolved and related to each other in the evolutionary tree. But the enthusiasm for embryology was lost in the early twentieth century when biology became dominated by the growing understanding of genetics. It is only recently that we have started to appreciate again how important an understanding of development is to the whole of biology.[14]

We now understand that environmental exposures during development are associated with choices that both predict and determine the environments to which we are best matched. Our biological processes are designed to respond to signals coming from the environment and to induce responses which can either be for immediate biological advantage (such as burning brown fat, a form of energy reserve in newborn babies, to generate heat when they are cold) or which help survival in the future. A grizzly bear puts on a lot of fat in the autumn so that it can use this stored energy to support body functions while it hibernates. The laying down of fat has no immediate advantage but is done because of the biological expectation of an impending winter. We refer to this type of biological expectation as 'prediction'[15] and our colleague Pat Bateson has used the term 'forecasting' in a similar fashion[16]—again, neither term implies any conscious intent. Animals do not gaze into crystal balls to foretell the future, but evolution has equipped them with the ability to gain information from their environment and use it to adjust their biology for future advantage. The grizzly bear's metabolic biology has evolved to be sensitive to shortening day length. And this type of biological forecasting even starts during fetal development.

So the developing organism uses information from its environment to make choices in an attempt to match its constitution to the environment it forecasts it will inhabit. In the same way, when we pack our bags for a journey we try to predict the weather we will face, and choose our clothes accordingly. If we have limited luggage, for example because we are travelling by air, we make choices about what to take and what to leave behind. Perhaps if we take an umbrella we will not take a raincoat. If we are going skiing, we will leave behind our shorts and sandals. Similarly, the embryo and fetus try to forecast their future environment and choose their luggage—in other words, they choose what kind of adaptive preparations and strategy will maximize their chance of reproductive success, because this is the ultimate goal of their biological journey. If they predict being born into a cold environment, they may develop a thicker coat of fur. This is what happens in the Pennsylvanian meadow vole, a small creature which looks like a cross between a mouse and a hamster. Like all voles, the meadow vole grows fast and mates within a few weeks of being born. If they are born in spring the pups have a thin coat of fur, but if they are born in autumn they have a much thicker coat. This difference in coat thickness is permanent and clearly has a survival advantage because winters of the north-eastern United States are frequently harsh even though the summers can be warm.[17] More dramatically, if an organism predicts that it will be living in an environment where there will be very many hungry predators, it may invest some of its developmental resources in defence. *Daphnia*, also known as the water flea because of the jerky way in which it swims, is a small freshwater crustacean that is popular with aquarium enthusiasts as fish food. In natural ponds, insect larvae are one of the main predators of *Daphnia*. But these larvae release chemical clues (called kairomones) which give away their presence in the pond, and if growing *Daphnia* detect high concentrations of these chemicals then they develop with a sort of body armour that makes them less vulnerable as prey.[18] In this book we will also see what happens if the organism's forecast is wrong, when an incorrect prediction leads to developing an inappropriate strategy, and how this leads to a mismatch. This will take us into the realm of a new science called ecological developmental biology (or 'eco-devo').

So how have we become adapted to the environments we inhabit? In part this is by natural selection, a term first introduced by Charles Darwin in his most important book: *On the Origin of Species by Means of Natural Selection* (1859).[19] This is the process by which organisms evolve through the selection of variations in characteristics, sometimes called *traits*, which make them more likely to survive and reproduce. Genes that influence the expression of biologically advantageous characteristics in individuals in a particular environment are more likely to be passed on to the next generation, because individuals possessing them are more likely to reproduce. So the mix of genes within a population, that is the amount of genetic variation in the 'gene pool', is changed over time and the characteristics of the species are gradually refined to match the environment.

But within a life course the environment also influences how the genes each individual has inherited from its mother and father are turned on and off. In the so-called 'plastic' phases of embryonic, fetal, and infant life environmental influences can mould how our characteristics develop, with permanent consequences. Evolution has equipped us with particular ways of responding to the environment during development and as a result there is a range of developmental choices and trade-offs which we can make; these are used to improve the chance of a match with the environment. Depending on the circumstances, the end result may be good or bad.

A match means two things are complementary. The shoes you are wearing match but are not identical, unless you have two left feet! When we say that two people in a relationship are well matched, we don't mean that they are very similar. In fact we may mean the opposite, that they understand one another well and that any shortcomings in the personality or behaviour of one are complemented by the other. When one is fed up or tired, the other will take action to help. The relationship works because it is dynamic, each person responding to the needs of the other in a mutually supportive way. In this book we are concerned with such a complementary relationship, that between the biology of humans on one hand and the nature of the environments in which we live on the other. Each is continuously changing to an extent and there is a constant dialogue between them. If the organism is *matched* to its environment, then we suppose that it has become suited by both evolutionary and developmental processes to be so.

The mismatch paradigm

A match does not have to be 'all or nothing'—there are degrees of match and degrees of mismatch, just as we might look in the mirror and ask ourselves how well a jacket matches the shirt we are wearing. The greater the degree of match between an organism's constitution and its environment, the more likely the organism is to thrive; the greater the degree of mismatch the more the organism has to adapt or cope. This incurs costs and these rise as the degree of mismatch increases. If the organism cannot cope at all, then the consequences will be a greater risk of disease or death. Mismatch can be created by changes in either the organism or its environment. It might arise for example as a result of a mutation in a gene essential for the organism's healthy life in that environment. The disease of lactose intolerance results from a mutation in the gene responsible for synthesizing the enzyme lactase. In the gut wall this enzyme breaks down the sugar lactose, found in foods such as cow's milk, making it possible to absorb it. If a person with lactose intolerance lives in an environment where cow's milk is a staple food, they will suffer from chronic diarrhoea from unabsorbed sugar remaining in the gut and they will also become malnourished.[20]

Alternatively mismatch can arise from the environment changing rapidly or drastically. Sailors in the eighteenth century had a high chance of dying on long voyages from scurvy, a very unpleasant condition in which bleeding occurs into many of the tissues of the body. The skin shows widespread bruising, gums bleed, teeth loosen, and recently healed wounds may break down. Bleeding into muscles and joints causes considerable pain. Scurvy developed because at sea the sailors' nutritional environment had changed from one where there was some fresh food to one in which all food had to be dried or salted. There were no fresh vegetables or fruit until they reached the next port, which might be months away, so their diet was very deficient in vitamin C. The human body stores some vitamin C and it takes time to run these stores down, but when they run out scurvy results. This is why the disease only became apparent on long voyages. But consuming only a small amount of fresh fruit could help. The sailors must have argued about this issue: how could fruit be kept for months in the warm, wet hold of a ship? The answer was to take vegetable extracts and then later lime and lemon juice on the voyage.[21] The sailors did not like the taste, preferring rum, but sauerkraut and limes saved the day, Britannia ruled the waves and Britons abroad became known as 'limeys'.

While so far we have used relatively extreme, but we hope instructive and interesting, examples to introduce the concept, much more subtle changes in our environment can also lead to a degree of mismatch and thus have very important consequences for human health. These will be the focus of this book. Many humans now live in environments to which they are not well matched and for many the degree of mismatch is increasing. Mismatch occurs in everyday existence in seemingly ordinary environments. And mismatch has a cost—it turns out to be a major determinant of our social structure, health, and quality of life. Understanding the *mismatch paradigm* gives us a new perspective on what we are.

PART I

Match

Over the next five chapters we discuss what makes a match and what can cause a mismatch. First, in Chapter 1 we will examine further the concept of match. In Chapter 2 we discuss the various types of inheritance that come together to determine our constitution, namely our evolutionary and genetic histories, the processes of epigenetic inheritance, and intergenerational behavioural and cultural influences. In Chapter 3 we move to a much shorter time scale to discuss how the processes of development adjust what we inherit in an attempt to make the best match with our contemporary environment. In Chapter 4 we think about the other side of the equation, by considering how the human environment has changed since our species first appeared about 150,000 years ago. In Chapter 5 we will show how together these various factors place constraints on how we can live in the modern world and how they generate situations where an unhealthy level of mismatch is more likely.

1
Our Comfort Zone

As we get older we develop a greater sense of mortality and we start reflecting on whether our life has been happy, has it been worthwhile, has it been successful? But what do we mean by worthwhile or successful? We might think somewhat hedonistically in terms of the possessions we have accumulated or we might consider longevity and quality of life as evidence of a successful life. Then we might think of some possible lasting impressions that we believe we have made on others. This is particularly true if we have children, and we wonder whether they reflect the values and ideals we have ourselves. Most of us who have children would in a sense see their existence and their achievements as the clearest evidence of our successful lives. Further, for many older members of society, a highly significant and gratifying moment in their lives occurs if they become grandparents.

When biologists think about animals and successful lives they use remarkably similar concepts. The most important questions for them are whether the organism has successfully reproduced and whether its progeny have in their turn lived to reproduce. Generally biologists use the word 'fitness' to describe this success.

Natural selection acts to select characteristics or traits that confer greater fitness within a given environment. In different environments different traits may be more advantageous. Thus after a change in vegetation produced by a change in climate, birds with blunter beaks rather than sharper beaks may have an advantage and be selected preferentially—indeed this is what has been observed in the studies by Rosemary and Peter Grant of the finches of the Galapagos.[1]

A basic principle of evolutionary biology is that evolution is driven by variation in characteristics within a population (that is, not all individuals are identical), and this variation is reflected in different levels of survival and reproductive success—so some individuals will have more offspring, others fewer. Provided that there are genetic components underlying their successful characteristics, those that have more offspring will enrich the gene pool of the next generation in this respect.

What must an animal do to have optimal fitness? The strategies needed will vary between species, between members of the species, and between the sexes. Much depends on the determinants of survival and reproduction. To win in the mating game,[2] the animal must survive long enough to mate, be healthy enough to mate, and, at least in birds and mammals, be able to nourish and nurture its offspring. There is no species in which all the progeny live to adulthood. Indeed in species such as fish and insects, a very low percentage does. In the green turtle, it is estimated that only 1 in 10,000 do. It is only in humans, whales, and a few other large mammals that the probability of a newborn surviving to adulthood is more than 25 per cent.

Optimal reproductive performance (maximal fitness) depends partly on the environment in which the individual lives—how scarce food is, how many predators there are, and indeed how many competing members of the species of the same sex there are. But reproductive performance also depends on the constitution of the organism itself—can it survive periods of food shortage better than its competitors, can it chase them away from its chosen mate? Optimal reproductive performance is achieved when the organism's constitution matches well the environment in which it is living. The processes of evolution will, over time, achieve this match as a species evolves. Not every individual in the species will be matched ideally as there is genetically and developmentally induced variation between individuals in any population. Indeed it is this process of matching, occurring for every member of the species over each generation as the species evolved, which has created most of the characteristics of that individual species. We say most, but by no means all characteristics, because some features may have arisen by chance, as a result of a mutation that confers neither advantage nor disadvantage on the individuals who possess it. For example, the distinctive coloration of some sea shells even though they live entirely buried in mud is likely to be such a neutral characteristic.[3]

And when we think about the features that contribute to sexual success, for either males or females, we realize that many of them don't have much to do with survival, fighting off predators, etc. Those species (like us) which have 'sexual reproduction'[4] generally select their mates to a greater or lesser degree, so this process of mate choice is also very important for the pattern of genes which are transmitted from one generation to the next. Before we leave for the time being the question of how the features of an individual arise, or have been selected through evolution, we must note that not everything that happens in biology has an adaptationist explanation—some characteristics may arise as the accidental by-products of some other unrelated adaptation or may not have any adaptational origin at all. For example, they may result from an ancestral mutation that is neither beneficial nor harmful.[5]

So a whole range of processes contribute to the reproductive fitness of an individual, some to do with the nature of the environment, others with the characteristics of the individual. Change the environment, and reproductive performance is likely to fall, a phenomenon well known to zookeepers who try to breed wild animals in captivity. But genes can also directly affect reproductive performance. For example in the merino breed of sheep there is a mutation called the Boroola mutation[6] which leads to these sheep having multiple ovulations, and sometimes quadruplets, quintuplets, or even sextuplets are conceived. Very few of these lambs survive as they are born small and the ewe cannot sustain enough lactation to support them on just two teats. On the other hand animal breeders know that inbreeding can often reduce fertility. In this chapter we will explore further how a match between characteristics and environment is achieved, and the range of possible consequences that follow if it is not. From the examples we have given it will be apparent that the degree of match, and thus the fate of an individual, depends in part on intrinsic factors and in part on the environment in which that individual lives.

What genes do

There are two terms which we now have to define—*phenotype* and *genotype*. They are important because they provide a useful shorthand which we will use throughout this book. Genotype simply means the repertoire of specific genes, including any small or large mutations, which an organism possesses.[7] All individuals in a species have a very similar genotype, although each gene within this genotype can vary in detail between individuals. There are two copies of almost every gene in every cell in the body, except in the sperm and egg which have only one copy. In all other cells, one of those copies comes from one parent and the other copy comes from the other parent. The other exception to the rule concerns the genes on the sex chromosomes. In mammals, the male has one X chromosome and one Y chromosome whereas females have two X chromosomes and no Y chromosome. The genes on the X and Y chromosomes are generally different.

Different versions of the same gene are called alleles, so for most genes we have two alleles—which may be the same or different, because one copy is inherited from each parent. The difference in the alleles is caused by differences in their DNA sequence. Such differences arise because the complex biochemical processes that maintain and copy DNA are not quite perfect. Every time a cell divides, the DNA making up the genes has to be duplicated by a series of enzymes which copy it, proof-read the copy and rectify any errors. Occasionally, errors in the copying process are not corrected and if present in the eggs and sperm they will be transmitted to the next generation. These errors are termed mutations. Some mutations have no obvious effects, others can cause major consequences for the individual. Some mutations are caused by changes in the sequence of the DNA for a single gene, as in the case of cystic fibrosis, whereas others involve major changes in the arrangement of DNA on the chromosome which will affect many genes, or even an extra or a deleted chromosome. Down's syndrome can be caused either by an extra copy of chromosome 21 or by extra chromosomal material from chromosome 21 on chromosome 14. Most alleles also have microvariation, much of which leads to only very subtle differences in gene function, if any, and can only be detected by determining the exact sequence of the DNA. These small variations are termed polymorphisms.[8] An analogy would be two editions of the same dictionary—one designed for the USA and one for the UK. In general, the words are the same but many show microvariation: color and colour, center and centre, plow and plough. If genetic terminology were applied to these dictionaries, we could consider the whole book as the genome and each word as a gene. We would say that the books were of the same species—they are genetically compatible, have the same genes, and we can form intelligible sentences using words with a mixture of US and UK spellings. However, they have different genotypes because when the genes are expressed they can produce different effects—equivalent to physical characteristics. We would note the presence of allelic variation caused by mutations which have major effects on meaning when expressed (for example, 'jumper' means a dress in the USA and a pullover in the UK) and by polymorphisms which have much smaller or no real effects (for example, replacing 'colour' by 'color' might irritate some British readers but it does not lead to confusion about meaning).

In contrast to the genotype, which we can never see unless we are in a laboratory and completely sequence all the genes (that is the genome) of an individual, *phenotype* describes the actual appearance of the organism. In common parlance it is generally used to describe an obvious physical feature: individuals may be described as having a tall or a short phenotype according to their stature. But the term phenotype can be used to describe any set of observable characteristics, even if to observe them we have to perform some form of diagnostic test. The biochemical phenotypes of a person with diabetes and one without are different, and this can be detected from the results of a glucose tolerance test. The cardiovascular phenotypes of a person with high blood pressure and of another with normal blood pressure are different. In general terms, the phenotype of an individual at any stage in life will be the outcome of repeated interactions between the genotype and the environment, from the moment of that individual's conception to the present. Because of this a particular genotype does not inevitably lead to a single phenotype. In many organisms it is this set of interactions during development that are critical determinants of phenotype and thus of fitness and survival. These are the processes which we call *developmental plasticity*.

Rather than specifying the phenotype directly, the genetic repertoire has given the developing organism the tools to allow its phenotypic characteristics to be moulded and matched to the environment, rather as musicians may perform a piece of music from a written score in different ways on different occasions, depending on how they feel, the nature of the audience, etc. Even organisms with identical genotypes, such as identical twins, may be very different in some phenotypic characteristics as a result of environmental influences—they are often different in their birth size because one has been less well nourished in relation to the other in their intrauterine existence. Recent research shows how they become progressively different in the patterns of gene expression as they grow older because they are subject to slightly different environments.[9]

The remarkable insights of Charles Darwin (1809–82) and Alfred Russell Wallace (1823–1913) caused a fundamental change in our understanding of how organisms interact with their environments.[10] In retrospect their achievement is even greater because they did not have any knowledge of how inheritance worked—the nature of genes was yet to be elucidated.[11] They did not know that genes are spirals of DNA which instruct the chemical machines inside cells how to make proteins. It is this genetic control of protein synthesis which is the basis of all biological processes including the amazing process of growth and development from a single fertilized egg to a mature adult made of 100 trillion cells organized in a very particular manner. A gene is a sequence of nucleotides (a form of molecule which combines a sugar bound to a nitrogenous base). Thousands of genes are packed end on end within a chromosome. There are only four different nucleotides used in DNA and yet genes are made of many thousands of nucleotides joined in specific combinations. There is a very sophisticated coding system of triplets of nucleotides (the so-called genetic code) to determine what each DNA sequence means, and thus which protein is made and when it is made. One end of a gene is not used to make the actual protein, so this portion is not 'translated'—rather it contains instructions about the conditions under which to turn the gene on or off. This part of the gene is called the promoter region. Other molecules, often proteins themselves, will bind to the DNA of this promoter region and so turn the gene on or off, thus starting or stopping manufacture of the protein. Much of the complexity of modern biology is about how the control of these regulatory factors or transcription factors (so called because they can induce the transcriptional activation or repression of genes by binding to their promoter region) can impact on the action of many other genes. So complex actions can follow from a change in the expression of one single gene.[12]

Variation

Selection is based on variation in appearance, structure, and/or function (i.e. phenotype) which, in turn must be partly dependent on variation in the genetic basis underpinning that phenotype. It was Darwin who saw for the first time the essential importance of variation.[13] Without variation there can be no selection—life would be all or nothing. If all the mice on an island were identical clones of each other (that is had absolutely identical genomes) and genetic variation never occurred there would be no further evolution possible for that species of mouse because selection would have nothing to choose between. Indeed it is inevitable that those mice would become extinct, because at some point there would be an infection or some other environmental change to which all the mice would be susceptible.

One of the most important components of genetic variation is that it confers differences in resistance to infection, so that not every member of the species is susceptible in an epidemic. Imagine whether the human species could survive the anticipated pandemic of the bird flu H5N1 virus if we all had identical genes. The flu virus is an expert at mutation which allows it to dodge the immune responses of its hosts which are finely tuned to defend against previously encountered versions of the virus. The reason why there is so much concern over the H5N1 variant is that humans do not yet have immunity to this particular strain. If the virus changes into a form that is easily passed from human to human, the seeds of the pandemic will be sown. Luckily (for some of us) we vary in our ability to mount effective responses to such a virus and not everyone who gets infected will be particularly ill. And fortunately for the human species we will develop immunity to the worst effects of this strain of flu, just as we have to all the others to which we have been exposed throughout history. If this were not so, we would have become extinct.

Every gene contains thousands of nucleotides and these can show variation. This leads in turn to variation in how genes are expressed and in some cases to the protein that is made. Because of this variation over generations the gene pool of a population may change, reflecting different mixes of this genetic microvariation in the individuals in the population and the frequency of these variations. As a result, some of the individual organisms within that population become quite different in their appearance (phenotype). As Darwin recognized, this can be accelerated by artificial selection. In animal or plant breeding, individuals are selected by breeders over many generations for features such as size, colour, or taste until some very extreme phenotypes quite distinct from the wild forms can be derived. Some of the fancy pigeons Darwin described looked very different from the wild pigeon dove from which they had originally been selected and bred. In the same way, the dachshund and the St Bernard might both be breeds of dog derived from a common ancestral wolf, but their relation to each other would not be immediately obvious to someone who had not seen a dog before. But breeders are also familiar with the fertility problems associated with some inbred strains. If the degree of shift in genomic make-up is so large that it makes the individuals no longer able to interbreed with the initial founder line or with the descendent populations, but they can nevertheless breed among themselves, then we have to say that a new species has been formed. This is one of the basic concepts of speciation[14] and one which was developed by Darwin.

In every species there are enormous variations in genotype which remain hidden. The variations in phenotype that such 'silent' genes can produce are only manifest under certain sets of circumstances, which can be uncovered by alterations in the processes of development.[15] There are important processes by which the influence of genetic variation is constrained, usually by other genes during development, to ensure that the organism is born conforming to the fittest design for the species. This process allows genetic variation to persist, rather than being eliminated, but not usually to be expressed. The latent variation remains in the genotype, and can be expressed under some circumstances—if the environment changes and a different pattern of phenotypic development is more adaptive. For example the plague locust *Schistocerca gregaria* can either stay in one locality if there is a lot of food available or it can migrate if there is a high population density and not enough food. The migratory form looks very different from the alternative solitary form—so different that if the switch between forms had not been observed, we would think that they must be two different species. The migratory form develops bigger wings, different mouthparts, a different camouflage colour, and a different metabolism so as to fly to where there is more food. It does not hide in the day but congregates in large devastating swarms which are not very discriminating about which plants they eat and leave devastation in their path. The signals for this choice of phenotype can come from the mother who secretes chemical signals about population density into the viscous plug around the eggs which she lays. When the larvae hatch they must eat their way to the surface as the eggs are laid underground. In addition, locusts are susceptible to chemical and tactile signals from other locusts which can influence which phenotype they develop. Thus the capacity to exhibit these alternative phenotypes is contained within the genetic information of the locust genome but is only expressed when food supplies are threatened.[16]

These processes are an important means of retaining latent genetic diversity on which evolution can subsequently act. They also tell us much about how developmental pathways are regulated and controlled. A most informative example comes from Siberia in the work of the great Russian geneticist

Dmitry Konstantinovich Belyaev (1917–85), who in 1958 started an extensive and important study of the Russian silver fox.[17] In the wild all these foxes have a silver coat colour and, as the word 'wild' suggests, they were usually not tame or friendly to humans. Belyaev noticed that there was however some variation even in this aspect of the phenotype: some foxes were tamer than others, and behaved more like domestic dogs when their handlers approached them. Belyaev was interested in why similar attributes of tameness appeared in a whole range of species as they were bred. So he classified a collection of foxes according to their level of tameness and bred only those that were the tamest. Sure enough, within a few generations his foxes began not only to be tame, but to look like pets: they had curly tails which they wagged, and floppy ears, and they even developed some attractive patterns of coat colour such as piebald, which would have been a very definite disadvantage in the wild. They also began to grow to different sizes and had different leg lengths, just like the varieties of domestic dog. Over eight generations the processes of artificial selection had led to the influence of a set of genes which held the code for developing the phenotype of a wild fox being diminished. At the same time a whole host of other traits, the genes for which must have been hidden in the genome but prevented from expressing their phenotypic traits, appeared.

Strategies for life

Every species has evolutionarily determined strategies for life. One set of strategies encompasses the key components of its life course: how it grows, how and when it reproduces, and how long it lives—these are the components that biologists refer to as its *life-history strategy*. Other components define how it lives within its environment: for example, what social structure it has, what it uses as a source of food, and what methods it uses to avoid predators. Each element of these strategies, and the interactions between them, has evolved by selection to optimize fitness. These life-history strategies are therefore intimately linked to the various environments in which the species evolved.

Organisms have evolved to use a range of reproductive strategies. The male Pacific salmon fights his way up the streams of Alaska, avoiding predators ranging from bears to eagles, and competes against other males to mate. After this he is exhausted. He breeds only once in his life and then he dies. Similarly the male praying mantis is eaten by his mate at the end of copulation, his final act being to provide food to support the development of the eggs he has just fertilized. One form of angler fish turns into a parasite on the larger female by biting into her, becoming nothing more than a parasitic testis living off the female's blood and dispersing sperm at the appropriate times. Such extraordinary examples of life-history strategies have always fascinated biologists. Darwin spent years of his life studying barnacles,[18] and specimens of an enormous number of species were sent to him by colleagues from all over the world. He noted that in some species of barnacle the males exists as a microscopic parasite within the female, having no independent life and never leaving the confines of her body—a very intimate relationship indeed!

But intriguing as these examples are, we need to remain focused on mammals. In mammals the female must nurture her young until they are independent because she is their sole source of food through lactation. Once they are mobile the mother, in some species assisted by the father, must teach the infant how to hunt or forage for food until they reach independence. In elephants and humans this takes many years. So a female mammal's ultimate success (or Darwinian fitness) depends on how many progeny she has, how well they are nourished, how many reach adulthood, and, most important of all, how successful they are in mating. Living longer will clearly allow her to have several pregnancies and so a greater chance of her offspring surviving. This is essential in species such as the sperm whale, elephant, or human where a single progeny is usually born after each pregnancy (technically termed a monotocous pregnancy) but it is also important in polytocous (multiple fetuses per pregnancy) and highly fecund species such the rabbit or rat because in these species the probability of an individual pup surviving is much less.

To female readers it may come as no surprise that the males of many species have very differing priorities and reproductive strategies from the females. In many herd species such as the impala or elephant seal, the male accumulates a harem and he fights with other males for dominance in order to retain this harem. Usually the largest male wins but he might hold his dominance only for one season. Even to achieve that single season of mating he must grow in strength, win in the battle for supremacy, and sustain it through the mating season. There is an alternative strategy that we can imagine: that is to be sneaky. We see it in operation in primates such as the gelada baboon. Adopting this strategy, a male remains small and insignificant. He avoids conflict with the bigger harem-owner but follows him around and strikes up a relationship with one of the females in his harem. This can allow him to mate with her surreptitiously.[19]

In other species the male lives in a stable coupling with a female for a prolonged period. This is seen in many birds, for example swans, golden eagles, and the wandering albatross, and in some mammals such as the beaver. Between these extremes are other strategies, e.g. several males mating with several females in a sort of 'free-love' arrangement as seen in lion prides, although the degree of 'fidelity' can vary enormously and there can be very elaborate hierarchies of mating rites. Clearly male fitness can be achieved by several strategies: multiple matings in a single season is one such strategy; long-term support of one or several females and their progeny is another.

Within the primate family virtually every social possibility and arrangement exists. At one extreme, the orangutan is a solitary species—adults have their own home territories and males and females only associate to mate. Offspring are brought up by the female with little input from the male. At the other extreme, baboons have adopted multi-male polygyny, living in large groups of adult males and females and their offspring. The males compete strongly for mating, and there is often a single dominant male, but they cooperate to defend their territory from other groups. Subdominant males may engage in 'sneaky' mating. Chimpanzees also live in multi-male, multi-female groups, but their social interactions are much more complex. Small subgroups, for example a female and her offspring or a male–female pair, may split off to search for food, later returning to the main group. There are diverse patterns of mating—for example, chimpanzee females may mate with several males, possibly as a strategy against infanticide by confusing paternity. Some primate species, for example the gibbon, are monogamous, forming long-term pairs of an adult male and female with their offspring. Gorillas form polygynous groups, with a single dominant male, possibly a few subordinate males, and a number of females with their offspring. The dominant male usually attempts to prevent 'his' females from mating with subordinate males or with solitary males living outside the group. Conversely, some New World primates such as marmosets adopt polyandry, living in groups of a single adult female with her offspring together with several adult males with which she mates. In this situation, although the males cannot 'know' that the offspring are theirs, their best strategy is to support the female and offspring on the assumption that this is the case. Finally, humans are generally monogamous, forming long-term pairs and with both parents having high investment in their children. However perusal of the more lurid Sunday newspapers will soon provide human examples of most of the strategies adopted by our fellow primates.

Because the human life-course strategy is based on a single child per pregnancy, a long period of nurturing of the offspring and a reasonably stable pairing arrangement between parents who have joint investment in the offspring is ideal. It is thus critical that the parents, particularly the mother, live long enough to support her youngest child until it is mature. In Darwinian terms this is designed to confer greater fitness in terms of passing genes on to the next generation. Fitness will be drastically reduced if she gives birth only a short time before she dies. We will discuss later the idea that this may partly explain why the menopause evolved. But we will also see the consequences of humans now living much longer than did our ancestors during our evolutionary history. Greater longevity means living healthily, being able to support one's offspring, and perhaps one's grand-offspring.

Apart from the threat from viral epidemics such as influenza, which are still very much a concern, humans are not at much risk from other species. But we have evolved in an environment in which threats from our own species are real. Warfare and intra-specific (human on human) violence remains a dominant issue in our survival, both as individuals and as a species (think 'nuclear winter'). Sociobiologists who examine behaviour from an evolutionary perspective (and we express some caution about how far one can go in applying evolutionary concepts to understanding human behaviour given our strong capacity to learn) suggest that we have developed a range of group behaviours to reduce the threat of competition between members of a human group. It is suggested that this is in part the evolutionary origin of behaviour such as altruism[20] and the development of our sense of moral and ethical standards.

An artificial distinction

We don't know whether the reader experiences the same sinking feeling as the authors when the phrase 'nature or nurture' comes up. We regard it as an artificial and unhelpful dichotomy—let us explain why. For many people nature means genes and nurture means environment. But this idea of nature as inherited genetic information encompasses a very great deal: indeed, as we shall see, not everything that is inherited is genetic. This alone shows the limitation of the approach. The genes that we have inherited are basically identical to the genes every other human has; what is different is subtle variations in these genes and in the way they are expressed.

Our DNA was duplicated from the DNA we inherited from our mother and father. In turn our parents' genes came from the duplication of the DNA in the single fertilized egg that made each of them, that is from the DNA of our grandparents, and that in turn came from our great-grandparents, and so on. About some 10,000 generations ago this takes us back to the dawn of our species. Speciation is a continuous process[21]—there was no magic single time at which our ancestors switched from being *Homo erectus* to *Homo sapiens*.[22] But we can keep following our replicating generations of DNA back through *Homo erectus*, to our early hominid[23] ancestors, to their precursors, to the beginning of mammalian life, and back earlier to invertebrate or even to single-celled organisms as our origins. Indeed this is what Richard Dawkins does so elegantly in his book *The Ancestor's Tale*.[24]

In every reproductive event there is a chance that errors in DNA replication will generate microvariation or even larger mutations. Some such changes are neutral, some are negative and are eliminated by natural selection, but some may be positive in a particular environment and so are magnified. Thus within a species we can see variation become stabilized within different environments. Over time organisms in isolation from each other (either physically or behaviourally) may accumulate differences in their genetic structure such that they cannot breed successfully with each other any more. Thus two new species can evolve from the same ancestral species. Humans and the apes, the gorilla, chimpanzee, bonobo (pygmy chimp), and orangutan, all evolved from a common ancestor. Our paths diverged from the ancestors of orangutans about 12 million years ago, from gorillas 7 million years ago, and from chimpanzees and bonobos about 5 million years ago—all relatively recently in evolutionary terms. We can use the similarities and differences in DNA to map our evolutionary history since life first appeared on this planet and we can also use it to map the relationships between groups of humans as they migrated out of Africa about 65,000 years ago and spread across other continents. It is not surprising that we share more than 95 per cent of our DNA sequence with our closest relative, the chimpanzee.[25] More surprising however is the fact that we share more than 40 per cent of our genome with fruit flies and 20 per cent with roundworms.[26]

This variation in gene structure can greatly affect the way we respond to our environment. For example the rare condition maple syrup urine disease arises from an underlying gene defect which makes it impossible to metabolize some amino acids (isoleucine, valine, leucine) properly. The excess amino acids and their inappropriate metabolites pass into the urine. The urine of the newborn baby smells like maple syrup—a rather unique way to a diagnosis. Untreated, this condition will soon lead to brain damage and death of the baby. Once diagnosed, if the baby is placed on a diet free of the offending amino acids it can grow up relatively normally. Unfortunately it is a very difficult diet and this is frustrating for the families and dangerous for the infant. But here is an example of a disease which could be considered genetic but at the same time is environmental, because when the child is placed on an appropriate diet there is no disease. The point we are trying to make is that it is entirely illogical to consider biology in dichotomous terms of genes and environment—all of biology is based on the continuous interaction of both.

These types of interaction are particularly evident with respect to environmental factors acting through development. How much calcium gets from the mother to her fetus to build its bones depends on her diet, her environment, and placental genes. If she has a low vitamin D level, either because of a poor diet or because she avoids sunlight,[27] the offspring will have brittle bones, but this brittleness is even more likely to occur if the fetal side of the placenta has a particular variation in the gene controlling transport of calcium across the placenta, leading to less calcium being available to the fetus.[28] So gene expression, behaviour, and diet all interact in determining how much calcium is deposited in the baby's bones. Such interactions between genes and environment are the basis of developmental plasticity.

Recent discoveries show that the very structure and activity of DNA itself can be altered within our lifetime by processes that we will describe in detail later in this book, themselves triggered by the environment. Suffice it to say here that the chemical structure of DNA can be modified at particular control points in its sequence, which can have enormous consequences for which genes are switched on, which off.

So the environment can influence the chemistry of DNA, not by changing the genes but changing whether, and by how much, they are expressed. These effects can be lifelong. So while it has become fashionable in biology to talk about gene–environment interactions the term becomes rather meaningless—just as does nature vs. nurture—when we examine it more closely. Because the environment may have altered the switches which form part of the genes, it is really very difficult to say what is gene and what is environment. The most important environmental signals that can induce this epigenetic change are those that occur in early development, and this is the first hint that our developmental history has an important part of creating our individuality. We hope that it is now clear why we believe that focusing our attention on more holistic concepts such as 'development' is more useful than perpetuating the debate about nature vs. nurture.[29]

Selecting for success

Unless one is interested in the role of one specific gene, it is helpful to think of the whole organism as one unit, one member of a species that interacts with its environment. It is that interaction that determines its fate.

One of the most debated questions in evolutionary biology in the twentieth century was whether the genotype or the phenotype is selected by the forces of evolution. At first, the answer seems obvious: the environmental forces act to select the phenotype, because this constitutes the characteristics of the individual and it is these upon which survival and reproductive success depend. But then on the other hand isn't it the genotype that is passed from generation to generation? Because selection results in useful characteristics being passed on to the next generation, won't it be the small variations in the genotype that lead to the diversity of forms upon which selection can act? So here we go again—getting embroiled in another nature vs. nurture, gene vs. environment debate. And, once again the truth is that both sides of the argument are correct. Genotypic variations produce a range of phenotypes and it is these that are selected by environmental pressures. To the extent that phenotypic characteristics are based on the genes, the phenotype 'stands in' as a surrogate for the genotype. This was the accepted dogma for most evolutionary biologists for most of the twentieth century.

But once we insert the dimension of development the inherent oversimplification contained in this dogma becomes apparent. It is not easy to understand how an undifferentiated fertilized egg knows how to differentiate into a mature organism, with all the characteristics of the adult. There was much debate over the extent to which this was genetically determined ('programmed') and to what extent environmental influences could modify the developmental trajectory. This problem stayed in the 'too hard' basket until two great thinkers, Ivan Ivanovich Schmalhausen (1884–1963) from Russia and Conrad Waddington (1905–75) from the UK, gave it an experimental and theoretical basis, but their important ideas were largely lost in the subsequent genomic knowledge explosion. For with the genomic revolution came the discovery of all sorts of ways of looking at genes in early development. The science of developmental biology became almost entirely focused on how genes control various aspects of embryonic development and the processes by which the cells of the early embryo divide and then become specialized, so that ultimately an organism is formed with over 200 different types of cell in distinct organs and systems. It has taken time to move from a purely genetic, programmatic view of development to a more complete view of how evolutionary principles and environmental effects interact during development. It has really only been in the last few years that a new, integrated understanding has emerged from the nexus between evolution, genetics, developmental biology, and ecology.

So far we have talked of the phenotype as if any particular characteristic was selected for in its mature form—to an extent this is true; the phenotype of brown eyes stays constant throughout life from when they first develop. But in many cases selection acts not on the characteristic itself but on the capacity to *change* in response to the environment, that is with the ability of the organism to adapt. This was a fundamental point to emerge from Schmalhausen, Waddington, and more recently from others studying developmental plasticity.[30] For example, if an animal lives in a fluctuating thermal environment, selection will act on its capacity to adapt and cope with that range of thermal environments. By analogy, you don't select a central heating system on the basis that it will generate a fixed amount of heat. A 10 kw heater might not give out enough heat in the winter but it will generate too much in the summer. You select a system that can give out variable heat and even reverse its cycle and cool, and so adjust heat output in relation to the environmental conditions.

Warm-blooded animals, including humans, cannot cope in very hot climates unless they can lose heat or are insulated from it and cannot come out on cold nights unless they can generate and conserve heat. Camels tolerate very hot days and cold nights in the desert better than a naked human could do because they have more extensive mechanisms to cope with extremes of hot and cold than we do. So during its evolution the characteristics of the camel have been selected on the basis of its ability to adapt to a broad range of thermal environments in the short term. But it cannot live for long in a very cold environment—it does not have the capacity to conserve heat in the cold except transiently overnight.

This concept that selection acts on the capacity to adapt leads directly to the idea of a 'comfort zone'—the environmental range over which the organism can adapt and still be reproductively fit. This zone does not have sharp boundaries—the more the organism moves from its optimal environment, the greater the cost it is likely to incur. The organism thrives best within the optimal part of the zone but may still be able to cope, albeit not thriving so well, at the boundaries. Indeed, many animals including humans move to the extremes of their comfort zone or even beyond it at least transiently. They take risks by going outside safe limits to compete with others, to find a mate and reproduce, to explore, or to escape from other threats. But in the longer term, living outside the comfort zone may mean that the animal cannot fully adapt—there will be a cost to being in such an environment and fitness may become compromised. Thus the Sherpa could live in an environment that created a greater risk of severe disease; but while many could adapt and live successfully in the steep mountain valleys, albeit at the price of developing a goitre, others could not and paid the price in severe developmental disruption to the brain. We can summarize our argument by saying that natural selection works to pick out the individual members of the species who have inherited, or who have developed, the appropriate ability to adapt and who can match their life course better to the environment they inhabit.

So much for the theory, but are there ways in which we can see it operating, or test it directly? There are indeed. The tiger snake is a particularly vicious reptile that lives in Australia and its off-shore islands. Snakes have jaws that have a kind of double hinge, allowing them to swallow small mammals, eggs, or birds whole. It turns out that some tiger snakes have big jaws, others have smaller jaws, but these big-jawed and small-jawed varieties live in different regions of Australia. The assumption for many years was that these were two different genetic strains of tiger snake, in the same way that giant poodles and toy poodles are different genetic strains of dogs which have different expression of the genes regulating the secretion of a hormone that regulates body growth.[31] So the logical explanation was that there were genetically different populations of tiger snakes with different expression of the genes controlling jaw growth. But clearly the different-sized jaws allow the snakes to eat different-sized prey. If tiger snakes lived in an environment where all their potential prey were large but their jaw size was small they would soon die out. In such an environment snakes with slightly larger jaws would be favoured by Darwinian selection and any genetic mutation that induced a slightly larger jaw size would be preferentially selected. Perhaps that is what had happened and that is why there were only large-jawed snakes where their prey were larger. On the other hand, if snakes lived in an environment where the prey were small, what advantage would there be in having a large jaw size? Growth uses energy and resources, and these would have been wasted in growing a larger jaw than necessary. In this environment, snakes with smaller jaws would be more efficient, and mutations that led to smaller jaws would be favoured.

The crucial role of early development

But recent experiments by Australian scientists[32] have shown that jaw size is not only determined by genetic factors but also by influences from the early environment. If the young from a population of small-jawed tiger snakes are given large prey, they grow up with large jaws. Here we can clearly see that the characteristics of the adult organism, and so its capacity to live in a specific environment, have been determined not only by genes but also in part by influences operating in its early environment. We suspect that future research will identify genes which determine the speed and size of jaw growth; or perhaps we will find that environmentally induced epigenetic changes in DNA structure perform this task. Either way, it means that some snakes are capable of altering their development to become big-jawed in an environment with large prey. Evolution has selected a tiger snake genome that conveys the ability to adapt to a range of prey sizes which this species encounters in the different environments it inhabits.

This is all very well, but does it apply to humans? It may do, because there are similar examples in human history. Malocclusion is a problem caused by having a lower jaw unsuited in shape and proportion to the upper jaw. It makes it hard to chew and is uncomfortable.[33] It is a relatively new phenomenon in the human—we do not find it in skeletons until after the seventeenth century. After that time it appears even in genetically stable populations which have not been changed by incoming migrants, suggesting that its appearance is not simply due to a new genetic variant turning up in the skeletal records. It is thought that the appearance of malocclusion is due to a change in infant diet from one which was coarse to the blended diet typical of modern infant foods. Because bone and muscle stay plastic throughout the period of growth and respond to the mechanical forces upon them, if there had been less chewing, there would be less stress and strain on the jaw and its growth would be impaired, leading to malocclusion. We were designed to have coarser diets than we do now. The price is in orthodontists' bills.

Developmental plasticity

Plasticity is a term used in biology to reflect flexibility in form and structure. Some tissues are plastic throughout life. For example, the size of many muscles can increase or decrease throughout the whole of life in response to changes in the extent to which they are exercised. This flexibility is contained within each muscle fibre. But other aspects of the body can only be altered during critical periods of development—for example the number of muscle fibres in the heart is determined in humans during fetal life and is not subject to further modification in later life. The process of flexibility which occurs in early life only during critical periods is known as developmental plasticity.

Developmental plasticity is a fundamental mechanism for adjusting an organism's characteristics during development to match its environment, particularly when an environmental change is more than very transient. It allows structures to develop differently and gene expression to be adjusted, in part by epigenetic means, in accord with the environment the organism senses during its development. Such processes can lead in some species to quite distinct forms of the same organism developing, even though they are genetically identical. This is called 'polyphenism'. This is very often seen in insects such as honey bees. They are worth considering a little further here, because there has been much research on them.

Honey bees have a very rigidly defined social structure which allocates tasks to different members of the hive. Worker bees and queen bees come from the same genetic stock and are both female, but which type of adult bee an individual develops into depends on how the larval form is fed. If fed the nutritious royal jelly made by special glands of the nurse bees in the hive, the larva will mature into a queen bee. She is able to reproduce, but has underdeveloped mouth parts and a relatively small brain. She also has large venom glands and squirts rectal fluid at her adversaries—not a friendly individual really. She does not work and is not adapted for food gathering. In contrast, the worker bee develops from a larva that is fed high-protein 'brood food' as well as some nectar and honey. A worker bee larva is also exposed to a different level of chemical signals (pheromones) from the queen. Workers have basket-like structures on the hindlegs for carrying pollen, a barbed sting likely to discourage predators and useful for hive defence, mouthparts suitable for foraging, and a large brain. But they are usually sterile. The workers travel enormous distances (estimated to total 50,000 miles and visits to more than one and a half million flowers to make a single jar of honey for our breakfast table) and retain detailed memories of the location of flowers and other landmarks. When they return to the hive they perform sophisticated dances to signal to other workers where they have been and what they have found. These dramatic differences between the adult types of female bee in the hive—the queens and the workers—are all triggered by exposure of the same female genotype to different types of nutrition and pheromones during development. They are essential to the survival of the honey bee in its environment and, because they are flexibly induced during development, they carry the ability to modify the numbers of queens and workers to meet changes in the environment from year to year.[34]

In other species, developmental plasticity may be less dramatic but nonetheless adjusts the phenotype to be matched to the environment in which the individual is developing. Unlike many other amphibians, spadefoot toads live in hot and arid regions. They breed in temporary ponds and survive the daytime heat of the rest of the year by digging burrows with their specially adapted hind feet which give them their name. Two species of spadefoot toad live in the same ponds in the Chihuahuan desert in Arizona. The way that the tadpoles develop from the egg in each species depends on the amount and type of nutrition available, but also on the numbers of tadpoles of the other species present. As they develop, the tadpoles grow mouthparts which are better suited to being either carnivorous or omnivorous, the former for eating each other, the latter for eating detritus in the pond. The carnivorous form (or morph) of one species is induced preferentially to the detritus-eating form of that species, especially under stress such as high population density, and vice versa for the other species.[35] This gives them both a very effective mutual survival strategy. If food is scarce and the tadpole numbers are high, this signals to them that they must change their development. The result is that the carnivorous tadpole form of one species increases and the omnivorous form of that species decreases. The opposite happens in the other species. Like Jack Spratt and his wife in the nursery rhyme, where he could only eat lean and she only fat, both spadefoot toad species have developed a strategy to ensure that their tadpoles get food.

Spadefoot toads are also interesting because the influence of their environment in early development does not end at the tadpole stage. Under optimal conditions in the pond, the tadpoles will develop to a certain size before they metamorphose to become toads, developing legs and lungs and losing their tails and gills. If the food supply in the pond is poor, or worse still if the pond starts to dry up, they must metamorphose early or die. Early metamorphosis is the best adaptive strategy under these conditions, but as so often it carries a price. Metamorphosing precociously means that the toads that leave the pond will be smaller. They are less able to compete with bigger toads from other ponds for food and for mates and, being smaller, they are easier for birds and snakes to catch and eat. Having successfully survived the immediate challenge of a poor environment during their tadpole phase, they now must face the price in terms of a tougher life as adult toads. The key point in this example is that the toad does not change its basic structure, as did the honey bee, but nonetheless responds to environmental signals during development in a way that has lifelong repercussions. The trade-off in body size has consequences for survival and on the capacity to reproduce. Such trade-offs are a common feature of developmental responses to environmental signals. It is an example of developmental plasticity of a type that has echoes in humans.

So while all biological processes have genetic processes underpinning them, the characteristics and the survival skills of the organism are not solely a function of its genes but the end result of how that organism has developed through sequential gene–environment interactions. Those interactions earlier in life influence how the whole organism will be able to respond to its environment at subsequent stages of its life course. These concepts are fundamental to an ongoing revolution in thinking about life-history theory and developmental biology. When considered in these terms it is not so surprising that early environmental influences, even if quite subtle, can have effects which can be magnified considerably later. This is especially true if the environmental changes occur in embryonic life, having major influences on the outcome of pregnancy and indeed on the rest of life.[36]

What is selected?

So the processes of selection include not just determinants of mature appearance but also of the capacity to adapt and, in a similar vein, the capacity to demonstrate flexibility or plasticity in response to environmental variation during development. These processes attempt to attain a good match between the phenotype of the individuals in a population and their environment, provided that environment is stable or varies in a predictable manner (e.g. with seasonal changes). Hence we see across the natural world all sorts of good matches between environments and the phenotypes of those organisms that inhabit them. Birds have beaks shaped perfectly for the type of food they eat. The oyster-catcher has a long sharp beak suited to probing in the sand, the vulture a blunt tough beak useful for tearing meat off a carcass. The famous different species and subspecies of finches of the Galapagos Islands have different beak shapes to cope with the different types of nut and seed that they eat. Other organisms have evolved characteristics to make it easier for them to collect their food or to keep them safe from predators: the polar bear is white to make it harder for the seal to see its approach, the stick insect has the appearance of a twig to make it harder for a hungry bird to find.

The fundamental issues which Schmalhausen and Waddington raised are yet to be fully integrated into current thinking, and there remains ignorance (even among biologists we regret to say) of the important differences between selection for defined phenotypic traits and selection for the capacity to adapt to environmental challenges which in turn leads to phenotypic change. Let us look at one further example to explain the problem. The rabbits of Australia all derive from a single stock released in the late eighteenth century to provide food for the early settlers from Britain. Now they are plague pests throughout Australia. The rabbits in northern Australia have longer ears than the rabbits of southern Australia.[37] This is adaptively logical because the ears are the major way by which rabbits lose heat and so longer ears are appropriate for a hotter climate. But did rabbits with genes for longer ears survive better as they migrated north? Was it selection of this trait that gave them greater fitness, so that any short-eared rabbits which migrated north perished? Or was it that the few rabbits which were imported had genes which conferred a capacity to adapt through the processes of developmental plasticity such that those born in warmer northern regions grew to have longer ears and those born in the colder south grew shorter ears? The tiger snake clearly provides an example of the latter form of induced phenotypic difference, but for the Australian rabbit we do not know. Natural selection can work over very short time frames[38] under some circumstances but it is not always clear what is being selected, a defined characteristic itself or the capacity to change it in order to adapt. To settle the issue we would need to take long-eared rabbits southwards, and short-eared rabbits northwards, to see what happens to the ear length of their progeny—whether changes in the length of the ears happen in a short time or over many generations, and whether the animals are born with different ear lengths or develop them according to where they are living.

Specialists and generalists

The examples we discussed above illustrate the power of evolution over time to match organisms to their environment. But these examples highlight another feature of the diversity of life. The better adapted to an environment an organism becomes, the more its life is likely to be confined to that environment. It may not do well outside that environment, because it does not have the appropriate adaptations. The anaconda does not have a phenotype that equips it to live in the tundra and the arctic skua could not survive in the Amazonian basin. Many organisms have evolved to live in very specific niches. The giant panda can only live in a narrow environmental niche between 1,200 and 3,300 metres in coniferous forests where the few species of bamboo it feeds on are found. We can only speculate on the selection process that put the panda in this situation, but the result is that its survival is at great risk as this bamboo becomes increasingly scarce. Similarly the koala can only live in parts of Australia where the blue gum tree grows. In many species of insects and invertebrates the environmental range for which they are adapted can be geographically very small. These are all examples of 'specialist' species where the degree of match between the environment and the organism is so close that the possible range of environments in which it can thrive is very small.

One way of thinking about this is that these specialist species are designed to live within very narrow comfort zones which are defined by the physiology of that animal—i.e. how well it can adjust its biology to meet variations that occur within the environmental zone—and the range and nature of such fluctuations in environmental conditions. So the optimal part of the comfort zone is not just determined by the environment but also depends on the adaptive ability of the species. The species is certainly able to cope with small changes in the environment, but a large unexpected change may be catastrophic.

But there are organisms that can somehow live in a very broad range of environments. Often we humans think of such organisms as pests, because instead of choosing to live outside, in what we see as their natural environment, they feel just as much at home with us. The common cockroach, rats, and mice are familiar examples—they can be the curse of households almost anywhere. Even the sly fox, so at home in the countryside, survives very well in many large cities by nocturnal scavenging in garbage bins. But arguably the most generalist species of all is *Homo sapiens*. Humans somehow manage to live in the Andes and near the Dead Sea, in the rainforest and in deserts. A baby has even been born to a woman working on the Antarctic peninsula. The skyscrapers of New York and the yurts of Mongolia—could one imagine more different environments? Yet to those who live there, both are very much home.

So far we have used the term 'environment' without being clear what it encompasses. It is not only the obvious physical environment (wet, dry, hot, cold, high or low altitude, mountain, plain, etc.) but the broader organic and social environment as well—the types of food available, the types and numbers of predators, competition with other species, population density, social structure and the capacity to find a mate, the burden of parasites, etc. All of these and many other environmental factors affect the capacity of that organism to grow, be healthy, and reproduce.

Just as environments vary, so does the capacity of organisms to adjust and respond to such environmental fluctuations. There is always an extreme of environmental conditions which will exceed the adaptive capabilities of a particular species, but that extreme may be surprisingly well tolerated by another species. Generalist species such as humans have a broad capacity to adapt or cope over a range of environments but may not be so well equipped to live in a particular environment as a specialist species. But it is important to distinguish between *thriving* in an environment and *surviving* or coping in that environment. Trade-offs which can affect our health and reproduction may have to be made once we move away from the centre of our comfort zone. The greater the shift from the environment at the centre of the comfort zone, the greater are the changes in physiology and behaviour needed to cope, until at some point significant costs appear. A panda can live in London Zoo but it does not thrive there nearly as well as in the bamboo forests of China—the environment is certainly not conducive to reproduction, much to the frustration of the zookeepers (and probably also the panda). A polar bear can survive in a temperate zone but again does not thrive. Whilst one of the authors was able to work at over 4,200 metres in the Himalayas, it cannot be described as pleasant and even the Sherpa do not normally live as high as this. Yet the climber gasping his way up a slope on Everest beyond this altitude may look up to see bar-headed geese flying overhead. How do they do that? They are adapted for their specific environment and have highly efficient lungs, flight muscles, and oxygen-carrying mechanisms in the blood which have evolved as a specialization to such a lofty environment. The deserts of the south-western USA are home to the kangaroo rat, a seed-eating rodent that hops on two legs just like its much larger namesake, although it is not in fact a marsupial. The kangaroo rat is so highly adapted (and selected) for life in waterless conditions that it requires no drinking water at all.[39] Normal carbohydrate metabolism generates both carbon dioxide and water, so the kangaroo rat can survive on its metabolically produced water provided that it has food. But humans cannot live in deserts unless they have access to water. So while we can cope in many environments successfully, we are not highly matched to some very specific extreme niches where other specialist species can thrive.

Being adapted

Living successfully for humans as for other species means being well matched to the environment—that is, living in the comfort zone. Humans largely deal with the physical environment by modifying it. We are not the only species to do so. The termite's mound is intricate and designed to control temperature inside despite wide fluctuations outside—termites are brilliant air conditioning engineers. And the beaver's lodge is well insulated to provide a warm home in the winter and to protect against predators. So does this mean that humans, like these other species, are well adapted to their environment?

To answer this question we need to be clear about what we mean by 'adaptation'. Evolutionary biologists use it strictly to refer to the results of selection which led to a match between the organism's constitution and its environment.[40] But it can also be used by physiologists to include very short-term adaptive responses, such as the processes of control of our internal environment (technically termed homeostasis): we sweat when we are hot to lose heat, we reduce urine output and become thirsty when dehydrated, and so on. Bringing the two extremes of time scale together, adaptation can be seen as the result of evolutionary processes operating over many, many generations to select the genetic determinants of the life-history strategy, the mature phenotype, the adaptive and homeostatic capacity, and the processes of developmental plasticity—each designed to improve the degree of match between the organism and its environment.

Much environmental change occurs in an unpredictable manner, or over a shorter time frame than evolution can deal with. In these circumstances the organism must try to cope with the change as best it can. It may be equipped with a number of structural and physiological devices that help to match it to the environment. So it may succeed to a greater or lesser extent depending on the degree of environmental change and its adaptive capability. Alternatively it may have to migrate to an environment to which it is better matched. Or it might attempt to change the environment. To address the question of how well humans are adapted to their environment we need to consider each of these strategies in turn.

Outside the comfort zone

Humans have a tremendous ability to modify their behaviour and their environment in order to cope in a wide variety of environments. But as we move further and further from the centre of our comfort zone, the potential costs rise. If we are lost in the mountains after a skiing accident and cannot put on extra clothes, we try to keep moving in order to keep warm. This will mean that we will expend more energy and will need food and water more quickly than if we had been able to keep still. An alternative strategy would be to dig a snow hole to shelter from the elements. This could be the safest plan at night, but it means that we do not attempt to get back to safety at that time and may not be so visible to rescuers. Calculating the cost of adopting a particular coping strategy under such circumstances can be quite difficult, and making the wrong choice may be fatal.

Sometimes we cannot fully adapt and this brings a cost. Goitre is the adaptive response the Sherpa made in an attempt to return thyroid hormone status to normal. In some cases it worked at the relatively minor cost of leaving the individual with a disfigured neck. In other cases the adaptive response was inadequate, with the serious consequences of low thyroid hormone levels and cretinism. Our research showed even more subtle adaptations that explain why some Sherpa were not affected by the symptoms of hypothyroidism. Thyroxine, the main thyroid hormone, comprises one molecule of an amino acid joined to four atoms of iodine. This is the main form that circulates in blood, but the active form of thyroid hormone only has three atoms of iodine attached—the fourth atom being removed in the body's tissues before the hormone is activated. But we found that in the Sherpa the big goitrous thyroid gland made the three-iodine-atom form of thyroid hormone in preference to the normal four-atom form.[41] This meant that for a given intake of iodine the gland could make more active hormone and secrete it directly into the bloodstream. For some individuals this was adequate to prevent the symptoms of hypothyroidism developing. This is a form of intrinsic adaptation by the body in trying to cope with iodine insufficiency in the diet.

When humans migrated progressively northwards from their ancestral home in the African savannah, they moved into colder climates where the average daily amount of sunlight fell. Occupying these new habitats in Europe and even further north gave advantages in terms of opportunities to hunt, and later to cultivate some simple crops and domesticate animals. But it also brought new threats. For example the low exposure to sunlight, especially in the winter months, reduced the production of vitamin D, which is made in the skin by the action of sunlight which converts a precursor molecule found in our diet into the active form of vitamin D. Vitamin D is vital for many body processes, notably the deposition of bone during development. People who have chronically low levels of vitamin D during development suffer from rickets, and this is associated with skeletal deformity. Older adults who are vitamin D deficient are more likely to suffer brittle bone disease (osteoporosis) and even minor accidents produce fractures. This had not been a problem in Africa as sunlight levels were high throughout the year, and our ancestors had evolved to have dark skins as the melanin protected against the other harmful effects of sunlight. In moving north we needed to evolve to have paler skins, filtering out less of the sun's rays and optimizing our production of vitamin D. The cost of this strategy is that there is a higher risk of skin cancer in people with paler skins, triggered even in Europe during the summer when the sunlight exposure can still be high. We do not need to extend this discussion further here, but it illustrates the point that humans as well as other animals have to trade off the advantage of one strategy—opting to live in a colder climate—against the possible longer-term costs of that strategy—behavioural changes in lifestyle in the winter and a higher risk of skin cancer. Such trade-offs have formed an important way of thinking about human ageing and we will return to them later.

Move or improve

One set of ways in which organisms can respond to changes in their environment, if these changes take them out of their comfort zone, is to change that environment. The most obvious way this happens is through migration. We usually think of this in terms of the feats of migration that some birds demonstrate: in following the summer the arctic tern migrates over 35,000 km a year between the Arctic and the Antarctic. But some mammals migrate too. The humpback whale migrates down from the southern Indian Ocean to spend the Antarctic summer feeding on small crustaceans called krill. In the winter they swim over 2,400 km north to mate or give birth. But there are also more subtle, and slower shifts which species make. With climate change the geographical zone in which the ponderosa pine in New Mexico can thrive is shifting.[42] In turn this affects the distribution of insects and birds that have evolved to be matched to the distribution of these plants. The study of these ecological changes is a major way of defining global changes in climate. For example the basking shark is thought to be moving its habitat northward as rising sea temperatures change the levels of the plankton which it eats.

Humans are excellent migrators. Much of this has been produced by a change in the environment—overcrowding and limited food supply is thought to have driven much of the great Polynesian migration. And the distribution of *Homo sapiens* and *Homo neanderthalensis* in Europe between 60,000 and 30,000 years ago was driven by the changing glaciation of the last Ice Age. Even modern-day nomads such as the Tuareg move continually to allow their livestock access to food, permitting them to survive in marginal environments.

But for humans there is a further strategy that has been even more important: why move to a different place when we can change the environment without moving? The development of fire, clothing, housing, and hunting tools were all hominid technological advancements which made it possible to live in a broader range of environments, and indeed to thrive in them.[43] Without these developments, the Inuit could not live in the Arctic regions. It is true that other species can also manipulate and create their own environments to a certain degree. The honey bee and the termite create their own environments in which to live and breed. Some termites have even developed a form of agriculture in which they keep aphids to milk as a way of generating their own in-house food supply. These specialized adaptations, from physical structure through to behaviour, enable different species to occupy different niches in the world.[44]

The reason humans have been such a successful generalist species is in large part due to our skill at technical innovation, allowing us to create housing and access to food even in disparate and changing environments. Humans continue to live in the Sahel even though it has changed from savannah to semi-desert in a relatively short time, particularly over the last few hundred years due to colonial over-farming. Another form of environmental manipulation humans have used involves the range of social structures we possess. These changed dramatically when agriculture was introduced because it required living in settlements, the development of specialized skills and of trade with other communities.

So humans have changed their social structures and other aspects of their lives to permit them to live in an enormous range of environments. But has there been a price for having this innovative capacity? What has been the cost of these changes for us? At first sight the answer might be: not much, because humans have done much more than just eke out an existence in their environments. In many of these environments they have thrived. We are after all a highly successful species. But perhaps the cost, the payback for successful adaptedness has been covert—this will be the subject of Part II of this book.

Failure

A very gross mismatch may be totally beyond the capacity of an individual to address. If it cannot address it and cannot move to a more favourable environment, and if this problem is also faced by other members of its species, then the future looks bleak—they face extinction. The fossil record which so inspired the Victorian geologists, and which was key to Darwin's development of the concept of the origin of species by natural selection, is full of evidence of species that no longer exist. The best-known and most dramatic examples of extinction, such as the dinosaurs at the close of the Cretaceous period (about 65 million years ago), were central to this thinking. The cause of the dinosaur extinction is not known, but a massive asteroid collision with the Earth is the most likely explanation. The more gradual geological changes of glaciation and uplift were crucial to the ideas of Darwin and his contemporaries and originated in the gradualist thinking of the great geologist Lyell.[45] Indeed Darwin wrote a book on the origin of coral reefs. Darwin had recognized the capacity of such changes to isolate groups of animals which would create a condition in which selection could act differently in different populations and lead to divergence and eventually to the generation of new species.

Any case of extinction on however small a scale, e.g. the loss of one species of tiny invertebrate, can be seen as an example of a point in time when the adaptive capabilities of that species were exceeded by the demands placed on it by its environment. Sometimes the environmental change may have been very slight, and it may have been simply the combination of a series of pieces of 'bad luck' for the species—a good season for breeding among its predators, the dominance of rivals for a food source, a small fall in temperature with poor growth of a plant food, etc. The effect for that species was no less dramatic than the effect on the dinosaurs of the asteroid impact on the Yucatan peninsula, which changed the face of much of the globe in a very short period of time.

The speed and timing of the evolutionary changes—loss of one species and origin of another—has been studied in great detail by those examining fossil and other records. It led to an acrimonious dispute between those believing in a gradual process of evolution and those who argued that it occurred in rapid shifts separated by longer periods of quiescence.[46] We are not going to join this debate here, but in any case we do not have to go back to the fossil record to find examples of extinction. The dodo was a large and plump flightless bird which lived in large numbers on the island of Mauritius. The island was uninhabited until the Portuguese and the Dutch arrived in the sixteenth century. The dodo had no fear of humans, in fact it appeared to be intrigued by them and the birds made little attempt to run from the sailors who eventually hunted them to extinction—indeed, they mistook its curiosity for stupidity and its name is derived from the Portuguese for stupidity. The extinction of the dodo was accelerated by the impact of other animals which were introduced such as the dog. Within eighty-three years of being first discovered, the dodo was extinct. The recent loss of many other species (more than 800 in the last 500 years, according to the World Conservation Union, and that is likely to be an underestimate) has much to do with an environmental effect—namely, the spread of humans to all parts of the world. The consequent changes in habitat are dramatic, due to our overexploitation of resources, introduction of invasive alien species, and climate change. Present rates of extinction are about a thousand times higher than the 'natural' rate estimated from the fossil record. This makes the effect of humans on other species comparable to that of the asteroid collision that led to the disappearance of the dinosaurs 65 million years ago.[47]

Extinction it seems has been no less a part of our evolutionary history. We now know that the migration of *Homo sapiens* from Africa 65,000 years ago brought us into contact in Europe with members of Neanderthal species, who also took their origin from *Homo erectus* but who had migrated earlier to Europe and had apparently thrived there.[48] We can imagine the consequences when the Neanderthals, who were so firmly established in their environment, met members of *Homo sapiens* with their greater innovative skills and ability to make sophisticated tools. So far as we know these two primitive hominids did not (or could not) interbreed. Their competition for food and shelter sites echoes many other scenarios throughout evolutionary history. The end result was the extinction of the Neanderthals.

The extinction of a species is obviously a biological failure for that species but the other alternative, moving to another environmental niche, may not be an option particularly for a specialist species. A more favourable niche may already be fully occupied by species or variants that are highly adapted and successful. Specialism may carry with it an in-built risk or hidden cost, because it limits the ability of a population to move outside its niche. Taking one adaptive path on a journey through place and time may rule out the option of taking other paths later.

Where are we?

Our environments have changed in many ways since humans first evolved. Are we just coping or adapting, and if so at what costs? Like other organisms, when we are no longer matched to our environment we must either adapt or pay a price by coping if we want to avoid extinction. That is the case in the goitrous Sherpa women who could still reproduce but unfortunately often gave birth to cretins. They were left to live life coping with their environmentally induced neurological defect. The Malinke people of the Gambia have to cope with extremely cyclical weather conditions—part of the year is very wet and is associated with famine, and only part of the year is associated with good agricultural production. Not surprisingly the incidence of disease in the hungry season is much higher than in the harvest season. Further, Gambian women who are pregnant during the hungry season give birth to smaller babies, who in turn have a greater risk of infant morbidity and mortality. These babies do not live as long when they become adults.[49]

Thus in humans, as in other species, there is an inherent cost to having a constitution designed for one range of environments and yet living in another. In many cases coping in humans is manifest as illness. Such mismatch is a major cause of much ill health and we will return to this concept later in the book. Sadly the cost of living in an environment, including a partially man-made one, mismatched to human biology is paid every day in both the developed and developing world in many thousands of lives.

2
Where Have We Come From?

It is always an enjoyable experience to wander into a pub in the Sussex Downs: it provides a friendly welcome, a place to relax, drink a beer, and engage in conversation with the locals. Once, we met an old farmer sitting by the fire, his dog snoozing peacefully at his feet. After the usual comments about the weather (what have we inherited to make this the inevitable first topic in any social interaction with strangers?), the conversation turned inevitably to what each of us did and why we were there. The farmer replied, 'Why am I here? I'm here because I live in the village and the other pub down the road is noisy and serves a lousy pint of beer.' This wasn't quite the answer we expected but it served to move the conversation on. Equally his reply might have been, 'What do you mean, *why* am I here? I was born here, I've lived here all my life and I'll probably die here. Why would I want to be anywhere else?' But he could have also replied, 'Ah well you see, my ancestors came over with the Normans so my family has lived hereabouts for nearly a thousand years.'

The farmer's three answers are all equally valid but refer to very different aspects of his journey to that particular pub at the point when we met. In referring to his ancestors, he was alluding to a long-term history of the people and events that led up to his life. In saying that he grew up in the village, he was referring to his own personal history. And in his comments about the pub down the road, he was explaining his presence in terms of some short-term environmental choice he has made.

Inheritances

Our experience in the Sussex pub illustrates the point that we can look at ourselves from very different time perspectives. Our DNA had its origin some 3.8 billion years ago in some crude primordial chemical mix that started replicating and has replicated ever since. It is the ultimate ancestor of all living things and has continued in an unbroken line ever since.[1] It was the capacity to replicate DNA that was the starting point of the continuum of life. The forces of evolution moulded the information contained within the replicating DNA to lead to the millions of current and extinct species of plants, algae, invertebrate and vertebrate animals. The first mammal evolved some 250 million years ago and the first hominid some 6–7 million years ago. From this ancestor arose a plethora of descendent species and from one of these, archaic *Homo sapiens* eventually evolved about 150,000 years ago. Then around 65,000 years ago some of these people walked out of Africa, and began to colonize other parts of the world. About 10,000 to 5,000 years ago, depending on our geographical origin, our forebears started to settle and engage in agriculture.[2] But our grandparents were only born between 60 and 150 years ago (depending on our age) and our parents only 35 to 100 years ago. And the egg that ultimately made us was formed when our mothers were embryos and their ovaries were formed.

So, depending on the perspective we want to take, we can see ourselves as carrying almost 4 billion years of inheritance, or only a few decades of inheritance. And what we inherited is not only the replicating DNA that forms our genes refined by about 3.8 billion years of evolutionary selection but other forms of inheritance from our parents and grandparents and from the society that we live in, and these other forms do not directly involve our genes.

This raises the question of what we mean in biology by the word 'inheritance'. One loose definition would be that it concerns those characteristics that run in families. A very tight definition (which actually excludes some forms of inheritance noted above) would say it is the transmission of genetic information from a parent to daughter generation. Geneticists have long been interested in familial or clustered patterns of disease. By studying these they hope to find the genetic basis for many diseases. But although adult onset diabetes has an 'inherited' risk component in that it tends to run in families it does not have a strict genetic inheritance pattern. Obesity has an even weaker genetic basis, although there are some rare purely genetic causes of both it and diabetes. But the origins of much obesity and diabetes clearly involve other factors, some of which can also be 'inherited', albeit in a manner different from a purely genetic mechanism. Eating and exercise habits are often similar within a family and this can be reflected in a common environmental contribution to the development of obesity or the appearance of diabetes. Often the two interact: those with a genetic predisposition may be at greater risk of diabetes if they eat excess calories and do not exercise enough.

So sometimes a familial trend may have both a genetic and a non-genetic basis or no genetic basis at all. A cluster of cancer across generations of the population surrounding Chernobyl does not necessarily suggest common genetic inheritance because common exposure to a toxic environment could be involved. But it could have a secondary genetic component because radiation can cause mutations in the DNA of the sperm in men exposed to it, and so cause effects in at least the next generation.

There are other less immediately obvious ways in which environmental effects can be passed across generations. People living in the Afar highlands of Ethiopia have lived in very deprived conditions for generations but when they move, say by being airlifted to a developed environment, they bring with them some biological memory of the deprived environment they left behind. We can see this because they are more likely to get diabetes as a result of going through a rapid nutritional transition from a poor to a rich environment. This risk for diabetes is not simply because they have genes for diabetes—it cannot be that simple because their risk of getting diabetes was low when they lived in the Ethiopian highlands. Instead their nutritionally poor early environment has affected the way that genes operate when they are later exposed to a rich environment and this effect can sometimes be passed across generations. This phenomenon can be termed *epigenetic inheritance* and is one of the most intriguing stories of modern biology, albeit one with a highly controversial and politicized history.[3] We believe that the epigenetic revolution in biology will turn out to be as important as the genomic revolution epitomized by the sequencing of the human genome. Indeed it has potentially even more far-reaching implications for human medicine.

Another form of inheritance does not involve genes at all—it is purely cultural. Many types of behaviour in a family can be transmitted from one generation to another. For example one of us has a son who has the bizarre habit of chewing bits of newspaper then spitting paper balls across the room—so did his late grandfather! Is this a genetic trait, or just a boy imitating his grandfather? How much of the current epidemic of obesity has its origins in parents establishing dietary and activity (i.e. lack of exercise) patterns in their children? This non-genetic but nonetheless inherited risk of disease is not really different from our preferences for certain sports, our religious beliefs, and our political views, which can be influenced in positive and negative ways by those of our parents. Few people would say that there are genes for cricket, for Jainism, or, thankfully, for fascism.

Much of the focus of medical research in the last fifty years, and particularly over the last decade, has been on the repeatedly stated dogma of genetic causation as *the* basis of most disease. But it is not always that clear-cut—much of what happens to humans in health or disease has cultural, developmental, and epigenetic components, and some of these influences are definitely inherited. It is for these reasons that we are concerned by the limited perspective created by the continuing dominance of the genocentric view, which underestimates the role of environment and development as determinants of risk of disease. Knowledge of the sequence of the human genome (and from a research point of view equally importantly those of the mouse, fruit fly, and roundworm) is without a doubt an enormous intellectual and technical achievement that has fuelled a new biological knowledge explosion. It has certainly soaked up a great deal of research funding too. But it is not everything—far from it.

In this chapter we will look at the various types of inheritance, using the perspective of the very different time bases over which they operate, and discuss the processes underlying them.

How evolution works

Bishop Wilberforce was shocked by the popular inference taken from Darwin's *Origin of Species* that suggested that we were descended from apes. The echoes of the famous debate in Oxford between the Bishop and Thomas Huxley over the theory of evolution have reverberated many times since; in the 'monkey' trial in Tennessee in 1925 of a young teacher (John Thomas Scopes) for using a textbook in his biology class which mentioned evolution, to the Dover School Board trial in Pennsylvania in 2005 over the teaching of Intelligent Design as a scientific theory. Actually the Bishop missed the point: we did not descend from apes, we and apes and mice and rats and sharks and toads and cockroaches and mosquitoes and the sea-slug and the earthworm all share common ancestry in some primordial single-celled organism. Before that, its ancestors had emerged from even more primitive organisms which had the ability to replicate but which may not even have possessed DNA. The ultimate origins of life, if it is defined in terms of replicating DNA or RNA, remain speculative and rank with cosmological issues of the origin of the universe as one of the big unanswered questions of science—they may never be answered.

But if we make it easy for ourselves and use the first replicating DNA as the beginning of our continuous line of descent then the unit of our inheritance is the gene and the process of determining that inheritance is evolution. Evolution works by selecting those genes that lead to increased capacity to reproduce in the current environment at the expense of alternatives which do not. The key point is that there must be variation so that some organisms are more likely to reproduce successfully (that is pass their genes to the next generation) and others are less likely to do so. Thus the three fundamental tenets of Darwinian evolution are variation, selection, and inheritance.

There are two related but distinct forms of selection. Natural selection occurs when the genetically based characteristics of the individual give it a survival advantage in one particular environment: when it is well matched to that environment it will be more likely to reproduce and pass these genes on to its progeny; if it is not as well matched to its environment it is less likely to pass on these genes. As a result of the continual operation of this process, the gene pool in the population changes. The process is generally considered slow but need not always be so.[4] When change is slow it is partly because things other than genes influence the characteristic being selected, but also because many traits have multiple genetic influences—for example there are well over 100 genes involved in generating jaw shape.

There are less than 25,000 genes in the human genome but infinitely more complexity in how the body operates. Some of that complexity is induced by the complex network of interactions in which a number of gene products can interact in generating a characteristic such as jaw shape. Some of it is produced by the complexity of the regulatory machinery which turns genes on or off, or adjusts their level of activity in different circumstances, and some of it exists because genes can produce different protein products by mechanisms operating both at the level of DNA and in the complex processing of the protein products of gene expression. Key components are regulatory (or transcription) factors, themselves products of gene expression which regulate the action of other genes—a bit like the stops on an organ which, in various combinations, influence the sound made by the keys and pedals. So whilst genetic variation partly drives that component on which selection can work, other factors are equally or even more important in generating this variation in any particular characteristic: among the latter, developmental and environmental influences are critical.

If the acacia trees in the savannah are tall then those giraffes with genes associated with development of longer necks will be positively selected, as these giraffes can eat better, be healthier, and are more likely to reproduce while those with shorter necks are more likely to be undernourished and succumb to illness. This is the classical description of natural selection at work, here selecting giraffes with longer necks. It is important to note that the giraffe was only selected to have a long neck because it gave an advantage in the environment in which it lived. The okapi, which is phylogenetically very close to the giraffe, has a much shorter neck because the important feature of its environmental niche was not the height of the leaves in the trees but the supply of shrubs and grasses closer to the ground. We use the example of the giraffe because of its historical echo. Darwin wrote about it and one suspects that this was because earlier proto-evolutionists such as Lamarck had already done so.

Jean-Baptiste de Monet, Chevalier de Lamarck (1744–1829), was an original thinker, a botanist, taxonomist, and polymath of revolutionary France at the end of the eighteenth century. His ideas have often been misquoted—indeed the idea most associated with him was not his own. His originality was not recognized at the time, especially by his rivals, and he ended his life discredited and in relative obscurity. His ideas are most often simplified to the view that the stretching of the neck which the giraffe would undergo during its life as it tried to get the juicy leaves at the top of the tree, might make its neck slightly longer; this attribute would be passed to its offspring, whose necks would also be longer. And, if they stretched their necks further still, they would pass on this in turn to their offspring. This concept is termed 'the inheritance of acquired characteristics'.[5] There is absolutely no scientific support for this theory—but Lamarck was one of the first to attempt to explain how hereditary processes could play a role in generating the similarities and the differences between species. But while acquired characteristics cannot be inherited, there is considerable evidence that environmental conditions affect characteristics which can in turn pass information to the next generation about the environment. There is now an unfortunately named (neo-Lamarckian) but scientifically important concept[6] that environmental memory might pass across generations—but in a quite different and important way that we will soon discuss.

Before we leave the giraffe it is important to note that more recent observations have suggested that the basis of the animal's long neck might actually be a different form of selection—sexual selection.[7] Male giraffes have to compete for mating rights, as do the males of many species. They do so by fighting for supremacy in a process called clubbing. In these battles the competing males swing their heads at each other like clubs. Here the long neck makes the head a very dangerous weapon, like a hammerhead on the end of a long shaft. Sometimes these fights are lethal. The giraffe with the longer neck will have a better club and is more likely to win the supremacy battle or the battle to impress the female and hence to mate.

This second form of selection, sexual selection, was the subject of another of Darwin's very important books: *The Descent of Man and Selection in Relation to Sex* (1871) which is often overlooked because of his *magnum opus, On the Origin of Species by Means of Natural Selection* (1859), but in which many ideas which form the basis of current evolutionary understanding take root. Sexual selection has the same fundamental basis as natural selection—that favourable characteristics are more likely to be passed on to the next generation. But what determines whether they are transmitted depends on whether they make the individual attractive to members of the opposite sex or more dominant, and hence increase the chance of a successful mating.

In most species not every male has the same opportunity to reproduce, unlike the females. In some species like the giraffe the male that is able to mate establishes dominance over other males—so the characteristics of the males that establish that dominance become magnified over generations. Darwin spent much time worrying over this concept—why these traits occurred in males alone and were not manifest in the female. He did not have the advantage of knowing the chromosomal and hormonal basis of sex determination, which in mammals allows a gene on the Y chromosome to induce testis formation, testosterone production, and thus the male to develop characteristics different from the female. A good example is the magnificent antlers of the red deer stag (they do not grow in the hind), used in competitions between stags to determine mating dominance during the rutting season. Such selection is also the basis for the gender difference in body size seen in many species—the males have become larger because they have been selected for bigger size by winning the battle for the right to mate.

In some species such as the angler fish and the elephant seal the difference in size between the male and female can be enormous. Such size differences may also represent the result of selection processes concerned with competition between the sexes, because the life-course strategies of males and females can be different yet they may compete for limited resources.[8]

But in some species it is not male brute force that determines mating success but rather active choice by the female. Her choice may not be based on physical manifestations of his strength but on her judgement of his appearance. It is difficult not to be anthropomorphic in this discussion. Is she judging her potential mate on some aesthetic basis? It might for example be his coloration, as in many species of bird such as the mallard duck, or in fish like the guppy where it is the iridescent blotches on his flanks that she appraises. It need not even be a physical characteristic, but might be some behavioural activity such as the dance of the male great crested grebe, the collection of coloured objects in the nest built by the male bowerbird, or the song of the male canary that influences female choice. Surveys in magazines often seem to conclude that one of the attributes women most value in choosing a male partner is his sense of humour.

Alternatively a female may assess the appearance of the males as a surrogate for some characteristic that she perceives as giving value to her or her potential progeny. The best-known example of this is the length of the peacock's tail, the longer tail of the cock attracting the hen. It is believed that this evolved through sexual selection, but equally it can be seen how it helps the peahen to identify a strong male as one that has the physical strength to carry such a heavy tail around. He would appear more likely to sire strong progeny.[9]

Another classic example of sexual selection carried to extreme is the Irish elk. This remarkable animal was neither particularly Irish nor an elk—it lived throughout Eurasia and was the largest known species of deer, standing 2 metres high at the shoulder and with antlers 3.5 metres across. It died out about 10,000 years ago, after the end of the last glacial period. The reasons for its extinction are unknown, although predation by the expanding human population of that time is most likely—and with those massive antlers, evading hunters in a world of increasing vegetation growth might have been difficult. Its antlers look impressive but were most probably used as courtship displays to reflect the strength of the animal rather than as weapons, and so sexual selection would have acted to increase their size still further.[10]

Both male and female characteristics may be subject to selection pressure, depending on the extent to which the male chooses the female and the female chooses her mate. In primates there are a wide range of social structures—ranging from harem arrangements in the gelada baboon to very communal behaviours in the chimpanzee that would put a Californian swinger to shame. These differences are reflected in sexually selected characteristics. The dominant gelada baboon male is large compared to the non-dominant males in the troop. In the chacma baboon, the red colour of the female's perineum is a signal of sexual receptiveness and is likely to have been a sexually selected characteristic. Likewise, in the chimpanzee, the phallus and testes are very large relative to body size in other primates.[11]

Darwin suggested in *The Descent of Man* that many human features may have developed through sexual selection. Presumably the protuberant female breast is one such selected feature; the loss of hair over virtually all our bodies which distinguishes us from our primate cousins is generally considered another.[12] And there is a difference in average height between males and females in all human populations, which might indicate an echo of a mating system where there was competition among males for access to females. Perhaps it was the dominant male in Palaeolithic[13] clans that had primacy in mating opportunities—a feature that continued into more recent times in the *droit de seigneur* and in the harem-type arrangements of some potentates in the not too-distant past.

Our genetic legacy

The first question virtually every mother asks after her baby is born, at least if she does not already know from an ultrasound scan, is whether it's a boy or a girl. Unfortunately in many societies the answer to that question has implications for the baby well beyond the colour of its first set of clothes. They may include infanticide or assignment to a lower social status with a reduced investment in nutrition and education throughout childhood and adolescence.

But the answer to the question is not always clear-cut—occasionally the doctor cannot tell because the infant's genitalia are not clearly either male or female. On examination there may be a phallic structure that might either be an enlarged clitoris or an underdeveloped penis with the urinary opening misplaced. It may not be clear whether there is an incompletely fused scrotum or very enlarged and partially fused labia. There has been some major disturbance in the child's genital development, but what sex should it be? Either this is a chromosomal female exposed to excess amounts of the male hormone testosterone or it is a chromosomal male in which the male hormone testosterone is not acting properly.

When such cases occur, the parents naturally want to know much more: how has this happened to my baby? Is it due to some event which occurred during the pregnancy? Is it genetic? We now know much about the way in which the sex-determining mechanisms are activated and translated into the anatomical structures that are either male or female under normal circumstances. The male (penis and scrotum) and female (labia and clitoris) genitalia are derived from the same embryonic precursor and, in mammals, it is the action of hormones from the fetal testis which shifts the pattern of development from the default—female—pattern to the male pattern. So hormonal problems during development can result in this process being incomplete.

The Dominican Republic is part of the Island of Hispaniola, one of the first Caribbean islands colonized by Spain after the so-called 'discovery' of the New World in 1492. One of the most unusual forms of intersex occurs here in one extended family in which children are commonly born with ambiguous genitalia. They are assigned at birth the female gender. But at puberty they show marked growth of their 'clitoris' and it becomes penile in appearance and they develop the male pattern of pubic and facial hair. Such children have clearly been present in this population for many generations and their fate and identity are well understood. Indeed the society accepts these children and has adapted so that they have particular roles and dressing standards before and after puberty, when they effectively go through a spontaneous sex change. They suffer from a genetic disorder in which the chromosomally male fetus cannot make enough of the active form of testosterone to virilize properly *in utero*, although at puberty the massive increase in testosterone enables some further virilization.[14] This unusual form of intersex is obviously only of importance in genetic males but it is a 'recessive' gene, meaning that one copy of the abnormal gene must be inherited from the mother and another from the father for the condition to be manifest. Such problems are more likely to arise in isolated populations where inbreeding between relatives is more common.

Other rare diseases can involve a 'dominant' gene, where inheriting only one copy of the abnormal gene passes on the condition. For example Huntington's disease is a form of dementia that develops in middle age. It is one of the most tragic of diseases—the person is perfectly normal until their thirties or forties then they progressively become totally demented. They usually die prematurely but may have already passed the gene on to their children. And only one parent need carry the abnormal gene. The disease is due to an abnormal gene coding for a brain protein called, appropriately, huntington which forms plaques in the brain and kills brain cells. While there is as yet no treatment for the disease, knowing the gene involved allows a test to be developed which can identify people at risk, even before they are born. The test can be used to help make decisions about whether to have children, or to continue with a pregnancy.

But such single gene defects leading to disease are rare. Generally very complex hierarchies of genes work to control all essential body functions, because many genes are not simply just turned on or off but are regulated to be partially on or off or only turned on or off under particular conditions. Each body function is like the mixing board of the sound engineer, although with many more equalizers and tone control functions to play with. How these controls are set will determine each inherent bodily function according to the environment, just as the sound engineer will change the settings on the mixer depending on whether the band is playing in a stadium or in a cabaret bar. But like the sound engineer, we can only adjust the regulation of our internal controls within the limits of our system.

The DNA sequence making up each gene can show microvariation (or polymorphism). The basic functions of these variant genes remain the same but the variation leads to subtle changes in the regulation or action of the gene which nonetheless can produce significant variation in the phenotype. Just as all violins look basically similar and function in the same way, but the timbre of the sound produced by a Stradivarius and a standard instrument used for teaching students are quite different because of the subtleties of the materials used and the way the violin has been built.

Animal and plant breeders have exploited this innate genetic variation by selectively breeding strains of wheat to improve its yield, horses for size and strength to pull the plough, or, particularly in the eighteenth and nineteenth centuries, pigeons with remarkably different characteristics for show. Charles Darwin was intensely aware of the history and success of artificial selection,[15] and he spent much time studying both the domestication of plants and animals and the tricks of the pigeon fancier. He expounds in the earliest chapters of *The Origin of Species* on this as an introduction to his new concept of natural selection and also wrote an entire book on the subject[16] and experimented in the vegetable garden of his home, Down House in Kent. The concept of the gene had not yet been elucidated, nor was the molecular basis of inheritance understood. Given that he drew his major conclusions in the absence of this knowledge, his insights are even more remarkable.[17]

Unknown to Darwin, crucial experiments on the inheritance of characteristics had been performed in Brno by the schoolteacher-turned-monk Gregor Mendel and published in 1865. Mendel showed in the monastery garden that crossing peas with different characteristics, for example green or yellow, smooth or wrinkled skins, could lead to a variety of types of progeny. He made calculations to explain his findings, based on the theory that some characteristics were dominant over others but that they would be inherited as distinct and independent entities. Looking back on his work we can see that this can explain why some human characteristics are inherited in a dominant manner (e.g. Huntington's disease) and some in a recessive manner (like the Hispaniolic intersex syndrome). The mechanisms of inheritance can be much more complex, particularly because many characteristics are not controlled by a single gene and as there are many ways in which gene expression can be affected. But luckily the inherited traits on which Mendel made his observations are not complicated by such problems. Mendel's fundamental discoveries, which he made working with the simplest of means, have stood the test of time as the basis for genomic inheritance.

When Mendel's work was rediscovered in the early twentieth century, the realization grew that the process of natural selection was compatible with the concepts of modern genetics—an intellectual development that became known as the Modern Synthesis or neo-Darwinism.[18] Amazingly, all this happened before the physical structure of the gene was understood. But by 1950 it was known that genes were found on chromosomes that were made from DNA. Studies of the DNA in the nucleus of the germ cells (sperm and eggs) showed that it was copied every time a cell divided into two daughter cells and that the amount of the genetic material (chromatin) was divided into two equal parts when germ cells were formed. When the sperm and the egg combine at fertilization, the total amount of genetic material is restored to its full complement, present in other cells of the body. The fertilized egg is therefore then able to divide and to start the process of development into another individual.

The more that the implications of the concept of genes were explored, the more all-encompassing as an explanation of much of life did it become. In 1953 the structure of DNA, the molecular basis for genetic inheritance, was established. By the end of the century, the Human Genome Project had completed the first draft map of the entire human genome, and maps of the genomes of other species, from malaria parasites to cows, have quickly followed. This escalating knowledge has understandably led to a very genocentric view of biology. But, while very important in advancing our knowledge of biology, this has also had detrimental effects. One is that it has led to a loss of interest by much of the scientific community in other perspectives such as the role of the environment and the potential for non-genomic inheritance. The second, related problem was that the genocentric view offered the possibility of positive, man-made selection of desirable genes in a population and the exclusion of unwanted ones. The consequences of such a view and the support early understanding of modern genetics gave to the eugenics movement[19] are well known. Even now there is considerable concern about the application of such ideas, from the development of genetically modified crops to designer babies.

The silence of the genes

Not all the DNA in our chromosomes codes for genes or for the regulatory parts of genes. Some of this material represents mutated or duplicated DNA which no longer has the necessary sequences to be activated. Some of this extra material, sometimes called inappropriately 'junk' DNA, reflects another part of our evolutionary history—our exposure to viruses. Viruses can be made of either a strand of DNA or of RNA surrounded by a protein coat. They do not contain the machinery to replicate their own DNA or RNA or make their own proteins. Instead they invade a cell and borrow the cell's replicative machinery to do so. In this process they may occasionally become incorporated into the host's genome. These viral relics no longer produce disease because cells have developed ways of switching off the action of these viral genes. So they remain there, a record of a history of infection at various times in evolutionary history, presumably many in species that antedate hominid evolution, now no longer threatening and no longer remembered.

So cells have switches that can be used to turn off old viral genes that have been incorporated into their genome. And they use the same kind of switch to control their other genes in the process termed epigenetic regulation—and these processes are critical to development. One common form of these switches involves chemical modification of the DNA by adding a methyl group (three hydrogen atoms and a carbon atom, $-CH_3$) at specific sites in the DNA sequence.[20] Methylated genes cannot be activated provided the methylation is in the control region adjacent to the main part of the gene. The important point is that the sequence of DNA itself is not altered, merely the way in which it is 'read', the methylation acts just as whitewash hides a word scribbled on a wall—it hides the gene from the DNA reading machinery of the cell. These modifications to genes are called *epigenetic effects*.

While we have two copies of most of our genes, for some we only want one copy turned on. Both copies being active may mean too much of a gene product is produced. Normally only one copy of a gene for a specific growth factor is active from fetal life onwards, the other being turned off.[21] If there are two genes for this growth factor active in the body then there is overgrowth of the fetus; it is born with a syndrome called Wiedemann–Beckwith syndrome in which the child is excessively large, has low blood sugar because the pancreas has too many cells and makes too much insulin, and is at marked risk of developing certain cancers. In Wilms tumour, a kidney cancer of children that starts before birth, the double dose of the growth factor is restricted only to the kidney.[22] Fortunately the tumour is readily treatable in most cases.

When we want only one of the two copies of each gene in our cells to be functional we sometimes silence the paternal or, for other genes, the maternal copy.[23] This silencing also involves epigenetic gene inactivation processes and is called 'imprinting'. There are not that many imprinted genes—perhaps only 100—and many are involved in regulating growth and development. This suggests that in some way the processes of interaction between maternally and paternally derived genes is kept in balance to ensure the optimal development of the fetus and infant.[24] We now know that many genes in addition to those which are 'imprinted' are controlled by epigenetic processes.[25]

Epigenetic processes must be an essential part of multicellular life. All the different types of cells in our bodies have developed from a single fertilized egg and must therefore all have the same genetic information. Yet they have developed differently, with distinct sets of genes being switched on to enable different types of protein to be made. A heart muscle cell must make the contractile proteins which allow it to function. A pancreatic secretory cell does not make such proteins, but it must have the synthetic processes necessary to make insulin. The development of these cell types from the embryonic stem cells which gave rise to them depends on extremely precise switching on and off of genes that control their development. Once a parent cell becomes a skin cell, all its daughter cells are skin cells. The multiplication of cells must also be controlled by genes. Some cells continue to divide throughout life, for example the cells lining the intestines or the cells of the skin, whilst others, such as those of the heart and the brain, essentially cease to divide before we are born. If the genetic mechanisms controlling cell division go wrong this can lead to cancerous growth of the tissue.

So the switching on and off of genes permanently is a very important part both of our evolutionary history and of our development. But this is not a closed system. We have already introduced the idea of developmental plasticity—the concept that one genotype can produce a variety of phenotypes. This must mean that the pattern of gene expression during development itself shows variation. Some of the changes may be irreversible—a queen bee cannot turn into a worker bee—whereas other changes are transient. For example in fetal life we use a different gene to make haemoglobin from the one used after birth and this leads to a different form of haemoglobin in our red blood cells at different stages of our life cycle. This results in mother and fetus having haemoglobins with different strengths for capturing oxygen and because of this oxygen is more readily passed from the mother to her fetus across the placenta.

Environmental echoes

The close match between the features of animals and the environments they inhabit demonstrates how much the environment moulds evolution. Without environmental boundaries and shifts there would be no selection—no genetic variant would be favoured over another. Genetic characteristics are selected within an environment according to whether they are favourable to reproductive success or not. To this extent over a very long time the genetically based features of an animal reflect the environments it has evolved in. So the now extinct dwarf elephants of Flores Island, which stood only 120 cm tall, are thought to be a reflection of limited food availability.[26] Such animals could only stay healthy and reproduce if they remained small, so their adaptive strategy was to match their needs to their food supply. Genes that promoted smallness were therefore selected and over time the elephants became smaller and smaller until their body size matched the food supply. We do not know why these elephants became extinct; possibly it was because of hunting by humans or following the massive environmental change produced by a giant volcanic eruption on Flores about 12,000 years ago.

This discussion has been about the influence of the environment over a very long time frame and this clearly happens through the processes of evolution. But in many places the environment is much more variable. Could environmental influences at one point in time have an effect later, not over many generations as for evolution but over shorter periods of time, say one or a few generations? While such transgenerational effects now have a solid basis in biology, this is a recent perception and the issue of environmental influences on inheritance operating over a few generations has had a chequered and confused history.

By the early twentieth century the arguments in favour of the Darwinian position had taken hold. But we must not forget that, even in *The Origin of Species*, Darwin recognized the difficulty of determining how much of a given characteristic (he used the example of coat thickness in mammals) was determined by long-term natural selection, and how much by more direct, shorter-term effects of the environment. He did not go into the latter in detail—he did not know about genes let alone epigenetic processes—but he allowed that some form of Lamarckian mechanism might play a role.

In Russia in the mid-twentieth century the idea of the inheritance of acquired characteristics had enormous political attraction. The agricultural biologist Trofim Lysenko had come to the attention of the authorities for his allegedly pioneering work on the effects of overwintering on crop germination. Like many plants, crops such as wheat grow better in the spring if they have been exposed to frost for a period in the winter. Faced with the problems of food production for the growing economy, especially in

Siberia, Lysenko conducted experiments in which he artificially treated grain with cold before planting. The results were claimed to be dramatic, and his technique of vernalization[27] held the promise of massively increasing grain production, possibly even to the point of obtaining two good harvests per year. Lysenko became an influential figure in the Russian scientific academy, being made Head of the Lenin Institute for Agricultural Sciences in 1948. This brought him into contact with senior officials in the Communist Party. The position went to his head, and he made increasingly extravagant claims about the effects that this new theory of science could deliver if applied on a grand scale. He applied to Stalin for funds to conduct grandiose agricultural experiments and it was years before it became clear that these were a disastrous failure. But in the meantime the Party's support for Lysenko grew, in part because the concept that environment could alter the characteristics of organisms was so alluring. If this could be done with wheat, why not with people? Could not this new science be used to improve the capacities of the workers, in particular to make those on state farms as efficient as those in factories? As Lysenko's influence grew he denounced many of his former colleagues in science, claiming that they were not radical enough or philosophically correctly aligned. The brilliant evolutionary biologist Schmalhausen was dismissed, to teach in the provinces. Another important and innovative geneticist and plant breeder, Vavilov, died under mysterious circumstances while being deported. It was not until Stalin's regime was collapsing that the whole fraudulent affair came to an end, although not without much public attention and debate.

With the benefit of hindsight the real tragedy of the Lysenko affair was not just the exclusion of original thinkers such as Schmalhausen, who died in 1963. It was an even more destructive, and equally politically motivated, effect which in some ways was more long-lasting. Post-Second World War Europe was dominated by the Iron Curtain, which blocked scientific as well as cultural communication with the West. Any information that leaked from the USSR was treated with the greatest suspicion. After all, hadn't the Lysenko affair shown just how lax their scientific methodology was, and how politically manipulated it could be? The work of Schmalhausen and others did not reach the West until much later, and Conrad Waddington, whose work was closest to his, seemed to know little of it. Taken together, all these events suppressed a willingness in scientists to consider how environmental factors might impact on the genetic framework, or to address the question of whether environmental factors acting over a shorter period than the evolutionary time scale could be important.

As an organism is most plastic when it is developing it was logical to imagine that environmental factors would have their greatest impact in early development. But the science of developmental biology, embryology as it was called then, was not linked to that of evolutionary biology. Indeed developmental biology played hardly any role in the 'Modern Synthesis' between Darwinian ideas and modern genetics, and it is only in the last decade that both theoretical and experimental work have shown that a critical component of biology had been underestimated. Even today the significance of epigenetic biology is only starting to be understood within mainstream biomedical science.

Yellow flowers and yellow mice

Epigenetics is defined as the branch of biology that deals with the effects of external influences on gene expression.[28] Increasingly the term is restricted to those processes by which chemical modification of DNA's function occurs by mechanisms such as methylation, without changes in the sequence of the DNA itself. This is the intrinsic process used in development so that from one single fertilized cell, specialized cells with quite different characteristics form. One line of cells will become nerve cells, with one repertoire of genes turned on, and another line of cells will become gut cells with quite a different set of genes turned on (and others turned off), even though both types of cells have exactly the same genotype and have originated from the same 'totipotent' embryonic stem cells i.e. cells which can potentially develop into any cell type of the body. This process which is internally regulated is critical to the normal processes of development. Here chemical signals from other cells surrounding any cell inform it of what its role is and induce epigenetic change. But could it be that this system can be affected by signals from outside the embryo—that is could environmental signals operating in embryonic, fetal, and neonatal life affect gene expression essentially permanently through these processes of epigenesis? The answer is an unequivocal yes, and in the next chapter we will describe how epigenetic processes underpin much of developmental plasticity. Parenthetically, there is much interest in the role of environmentally induced epigenetic change as the basis of some cancers.[29]

We are concerned in this chapter with inheritance. Can epigenetic change be inherited? When one cell gives rise to daughter cells, as in the case of a growing organ, those daughter cells have the same characteristics as the parent cell. So daughter liver cell precursors have the same pattern of gene expression as their parent cells—the same genes are turned on and the same genes are turned off. In a sense this is a form of cellular inheritance and although the biochemistry of how this happens is still somewhat murky it is clear that the epigenetic profile of 'on and off' genetic switches must have been transferred from mother cell to daughter cell.

But what about between generations of individuals? Traditionally it has been assumed that the epigenetic 'marks' (or memories) that determine the pattern of gene expression are wiped out in the processes of fertilization and early formation of the embryo. Essentially the fertilized egg is the ultimate stem cell and has to have the capacity to replicate and develop into the myriad of cells that form the body. So it must have a fairly clean slate free of epigenetic marks. It is generally thought that epigenetic marks such as methylation are only re-imposed as the cells start to differentiate into their various types with different patterns of gene expression. But it appears not to be as simple as this. There are now compelling experimental data showing that some epigenetic marks, representing the influence of the environment on one generation, can be imparted to several subsequent generations although the underlying mechanisms are not yet fully understood.

The idea that environmental influences can be transmitted between generations is well recognized in both plant and animal science but is still new in medical science. The toadflax is a flowering plant in which the shape of the yellow flower can be in two distinct forms—one rather pretty and one unattractive. Plants with each type of flower tend to breed true. Indeed the great eighteenth-century taxonomist Carolus Linnaeus thought they were two distinct species. But every now and then one flower form can flip to the other. Molecular science has recently shown that these two types of toadflax are identical genetically—the only difference is that one gene, which controls petal symmetry, is active in the pretty form of the plant and inactive in the other. The mechanism involves an inherited epigenetic mark that turns off the gene in the less attractive form.[30]

And there are fascinating examples of these processes at work in mammals. The agouti mouse is an experimental mouse that can be born with a yellow or a brown coat colour. The amount of yellow depends on how much agouti protein is expressed from the agouti gene. The amount of agouti protein expressed is transmitted from parent to offspring, albeit imperfectly. Whether the gene is very active or less active is controlled by epigenetic processes—less methylation results in a lot of agouti protein being expressed. Intriguingly recent studies have shown that changing the diet of the mother mouse before conception and throughout pregnancy to modify the biochemical pathways controlling methylation will alter the degree of epigenetic control of the gene, and this is reflected in the coat colour of her offspring.[31] In other experiments, also in mice, administration of hormone disruptor agents found in pesticides interferes with sperm viability, numbers, and motility. Intriguingly, recent experiments show that even if the chemical is only administered in one generation, effects on the sperm can be seen in the third generation of progeny,[32] again due to a transmitted epigenetic change. These are perhaps rather abnormal situations, concerning a peculiar gene in one strain of mice or the results of exposure to rather high levels of toxic chemicals. But recent research shows that changes in the diet of the pregnant rat or stress hormonal exposure in pregnancy produce changes in the physiological control systems for metabolism and blood pressure control in the offspring and that these are mediated by epigenetic changes in gene expression which can also be transmitted to the third generation.[33] So there can be little doubt that an environmental influence on one generation can leave epigenetic marks which can pass on to subsequent generations.

But epigenetic processes are not the only way environmental influences can pass from generation to generation. When a mother is undernourished she gives birth to a small baby. If the baby is a girl, she may grow up to be more likely herself to give birth to a small baby. Indeed, this can happen if the undernutrition occurs only in the first part of pregnancy. Some of our knowledge of the effects of famine on human pregnancy and the subsequent health of the offspring comes from a tragic event during the Second World War known as the Dutch Hunger Winter. In the winter of 1944/5, food supply to part of the Netherlands was drastically curtailed by the occupying Nazis as a reprisal for resistance activities. The resistance had been particularly active at that time in the war in order to support Operation Market Garden, the disastrous attempt by the Allies to capture key bridges and transport links in the Netherlands. Part of the resistance effort involved a transport strike. When the operation failed, the Nazis banned transport of food to the western Netherlands by rail. For some time it was possible to get supplies through by canal, but the winter of 1944/5 was harsh and the canals froze. The famine lasted some months but was quickly relieved after liberation of the low countries by the Allies. Importantly, detailed medical records continued to be kept by some hospitals and it has been possible to follow the health of the affected population and their children over subsequent generations. One of the findings from such studies was that baby girls subjected to maternal undernutrition during the first trimester of their gestation could be born at normal size, but when they grew up and became mothers themselves they gave birth to smaller babies.[34] Moreover, studies from scientists in Spain suggest that girls born with lower birth weight have a smaller uterus, perhaps because the uterus largely forms in the first half of fetal life.[35] A smaller uterus can exert more constraint on fetal growth, meaning that a woman with a smaller uterus will give birth to smaller babies, giving another biological mechanism by which transgenerational effects can be transmitted.

Cultural legacies

But animals, particularly humans, have other ways of transmitting information from one generation to the next. They use the power of communication through language, teaching and learning, and behavioural training and mimicry. This third form of inheritance is sometimes—and perhaps controversially—called 'cultural inheritance'.[36] Cultural inheritance can involve attributes being passed from one generation to its offspring (in biblical terms from father to son, mother to daughter) or it may follow non-familial routes—between people who are not related genetically.

On the Japanese island of Koshima, there is a colony of macaque monkeys that have been extensively studied by scientists.[37] They are fed potatoes as a means of attracting them to sites where they can be studied. But the site is sandy and, just like human picnickers on a beach, the monkeys prefer to eat a potato that is not covered with sand. One matriarchal monkey, Imo, adopted the habit of washing the sand off her potato in a stream—so she enjoyed a sand-free snack while the other monkeys were eating their rather gritty potatoes. But now, several generations later, all the monkeys in the colony wash their potatoes. They have gradually learned this culinary trick from each other and passed on the tradition of washing potatoes to their offspring. The same colony has shown that they could learn more. When fed wheat grains this too got muddled up with sand. One monkey found that if she threw a handful of the mixture into the sea, the sand sank whilst the wheat floated and so could be scooped up and eaten. Now when fed wheat all the monkeys in the colony sieve it and wash it by throwing it into the sea. And so cultural change continues. Going into the sea to wash their food made the monkeys discover that it was fun splashing around there, and so a new macaque tradition sprang up. Some older monkeys started eating fish discarded into the sea by local fisherman. Now no longer afraid of the water, the local pastime for the macaque males is collecting fish from rock pools, so it isn't only in *Homo sapiens* that fishing is a favourite hobby of the male. Here we see the complexity of cultural inheritance at work—one specifically advantageous behaviour is passed to others including offspring, and is then elaborated upon such that complex additional behaviour emerges.

There are many such examples of learned skills being passed among colonies of animals. Chimpanzees use primitive tools to fish termites out of their mounds, but in different ways in different parts of Africa; some colonies of baboons will walk upright across a stream, others cross a stream on all four legs.[38] And birds adopt different nesting strategies as they learn how to avoid predators.

Cultural changes can spread fast. But the substrate must be there. The capacity to learn and copy must be present and this itself is genetically determined. Over time there may be natural selection as well, so that those animals most capable of learning are more likely to survive to breed. This we believe is what happened in hominid evolution. The capacity to learn and communicate became critical to the survival of the archaic hominids and gradually they developed bigger brains to support these higher functions. The development of the first tools by early *Homo* species about 2 million years ago and the use of fire perhaps 1 million years ago were important steps in our evolution. But the capacity to communicate in a highly sophisticated manner by the use of language was probably the most critical step. The brain pathways controlling speech and the formation of the larynx to make sounds were critical anatomical developments and they must have evolved by classical Darwinian processes.

But the nature of language and its use is largely determined by 'cultural evolution'. Many thousands of languages evolved and groups of humans in isolation from each other developed very distinct language forms. In New Guinea alone, where valleys were separated from each other by impassable mountains and tropical forest, there may have been well over a thousand languages. Human languages fall into obvious groups—German and English are related, French and Spanish are first cousins, Maori and Tahitian are similarly linked. And languages can evolve with very different emphases in different communities. In many aboriginal languages there are much more complex sets of terms for specific relationships within a family than are used in English, where we might use the term cousin to cover all sorts of complex intra-familial relationships. The aboriginal languages reflect the importance of the extended family unit in their societies and the various marriage taboos that can be operative. In Albanian there are more than twenty-five words used to describe different types of moustache[39]—yet the concept of a moustache on its own would be quite foreign to the Waghi from the New Guinea highlands for whom a full beard is a core component of their identity and who would not have use for such descriptors. A Romanian orphan adopted at birth to England grows up speaking English, not Romanian. These are all examples of cultural evolution and inheritance at work.

New languages can emerge even without speech. In several deaf communities people have worked out very sophisticated sign languages which allow a full range of expression to be communicated. One recently documented example is that of the Al-Sayyid Bedouin sign language. An extended inter-marrying Bedouin family has common descent from Al Sayyid, who settled in the Negev desert 200 years ago and carried a recessive gene for deafness. The first deaf children were born early in the twentieth century. They invented a basic sign language which in turn they passed on to their children, But 100 years later it is now a complex and sophisticated language used not only by the 80 deaf members of the community but by all 3,000 members of it. Importantly for the study of linguistics, it is quite distinct in its structure from the languages that surround it.[40]

Changing language and idiom is a clear example of cultural evolution at play as new words or meanings can enter a language. The language spoken on the outer Georgia Banks islands such as Ocracoke is derived from English but is quite distinct. Pidgin is an example of a new language that has appeared since Europeans invaded the Pacific and a common patois needed to be developed to allow communication. Within families some words can have a particular and coded meaning and professional jargon is not dissimilar. And even within one language different societies give words different meanings—a 'rubber' in Great Britain means an eraser, in the USA it means a condom—a use of words that has given many a traveller some cross-cultural embarrassment.

There is still much debate about how to interpret the archaeological record in relation to when language developed. Was it more than 200,000 years ago or as recently as 50,000 years ago?—there are different views amongst palaeontologists.[41] Sophisticated communication is necessary for creative activity such as art, which appeared between 50,000 and 70,000 years ago. It is generally thought that the Neanderthal had some limited language capacity as they lived in complex societies which involved some collaborative and coordinated activity. But it is in modern humans, following the great cultural explosion of some 30,000 years ago when symbolic art as we know it and cultural practices such as burial appeared, that language must have played an important role in driving our cultural evolution.[42]

Concepts that we would now term religious most probably appeared at about the same time. The evolution of religion is in itself a fascinating concept—although one beyond the scope of this book—but it again demonstrates how sophisticated concepts of belief and custom can be adopted by a group and establish very strong behavioural attributes. It has been suggested that religious practice created a set of rules that became particularly necessary to permit groups of individuals to live in larger colonies, and form more complex societies once agriculture and settlement were beginning to develop.[43]

Today we see this cultural inheritance demonstrated in the observations that religious and political preferences run in families. And social structures in different societies, whether they are matriarchal or hierarchal, control reproduction (from harems to polyandry). All these are determined by similar processes of cultural diffusion and inheritance. They generate intergenerational bonds which are very strong and which if broken can lead to a loss of continuity between one generation and the next. In orthodox Jewish families, the cultural requirement is that marriage is to another Jew. If a child breaks with that tradition, it may well lead to the parents considering their child as dead and conducting mourning prayers and rituals. In some societies, the horrific custom of honour killing reflects the strong commitment to cultural inheritance over biological inheritance, and the long-running ethnic feuds in many parts of the world show how difficult new generations find it to forget the wrongs done to their ancestors.

An important consideration in this book is the spread of agriculture. Agriculture developed independently in several parts of the world.[44] But once developed it spread, particularly in Eurasia from its site of origin in the fertile crescent of Mesopotamia. How did this occur? Did farmers migrate from one region to another, or did one tribe learn by observing a neighbouring tribe and copying it, thus spreading the technology—rather as intellectual property is transferred today on the internet. Agriculture and its accompaniments—with religion, social custom, and language—have been the most important parts of cultural evolution.

But even at a family level cultural inheritance is critical to health. For example, mothers teach their daughters how to care for infants. In rural communities such as in the Gambia, child survival is much higher in families where the grandmother is there to assist the mother.[45] And patterns of childcare are transmitted within communities. During the later part of the twentieth century there was a very high incidence of cot death, up to 8 per 1,000 live births in some places. It turned out that cot death was more common in those babies put to sleep on their stomachs, but at that time the community wisdom was that babies are happier and are less likely to regurgitate and inhale a feed if put to bed in this way. Mothers taught their daughters who then placed their own children on their tummies. Child health nurses and paediatricians did the same. Well-intentioned cultural inheritance had gone wrong. But once research showing that the risk was less if babies were put to sleep on their backs was published, it only took a few newspaper articles, television programmes, and an orchestrated information dispersal campaign to parents ('Back to Sleep' was the slogan) to change this cultural practice.[46] How much else about child development is influenced by cultural inheritance? Breast feeding practices in different cultures vary. Many practices in pregnancy, childbirth, and infant care are embedded in the cultural identity and practices of a community—yet all may have consequences for the health of the offspring.

3
When We Were Very Young

Newts and salamanders are fascinating creatures. Biologists love studying them and in aquaria they are favourites of children. The alpine newt which lives in lakes in southern France and Switzerland is primarily an aquatic creature—its eggs are laid and hatch in the shallows at the lake edge and the baby newt looks very much like a tadpole. And like tadpoles, they live entirely in the water. But as they grow they have a choice. They can stay as tadpoles and in that form they can be reproductively active, even though they remain infantile in appearance. But alternatively the gill slits can close over and these newts can live both in the water and on land. In any population of alpine newts both forms can be found although the proportion varies. But the lives of these two forms of the newt are very different. Interestingly it is usually the bigger infants that undergo metamorphosis and close their gill slits. But those that stay infantile, while they start life smaller, can swim deeper in the lakes, because they retain their aquatic gills. In deeper water there are more plankton and less competition, so as a result they end up growing faster and they have greater reproductive competence. But when times are tough and water levels fall, it is the newts that are capable of spending time on land which have a greater chance of surviving. Thus under some circumstances it is better to be a persistently aquatic newt, at other times it is good to be able to leave one drying pond and get to another.[1]

So development can be complicated: it is not just a simple linear and automatic 'programme' by which the fertilized egg grows into an adult using only the information carried in the genes. And it is not a closed process—the newt derives information from its surroundings about population density and competition for food and then adopts a particular developmental strategy. The result is a very different life course—different growth pattern, different food sources, different chances of survival, and different chances of reproductive success. The evolutionary capacity to make such choices in early development is found in every part of the animal kingdom from single-celled organisms to the human. In some cases the results of the choices are obvious—gill slits or no gill slits. In many others they are much more subtle, but in all cases they serve an underlying purpose—an attempt to match the organism to the environment it will inhabit so as to maximize its chances of reproductive success.

Thus we must discard the notion that the genome is like a perfect blueprint which lays down a set of instructions so that the fertilized egg starts a series of cell divisions and differentiation until the organism reaches its preordained mature state. If that were the case then every individual with the same genotype would be virtually identical, and that is clearly not the case. Even identical twins are never identical in every respect—for example they are generally of different sizes at birth. The blueprint analogy fails—yet it is amazing how often it has been used. Instead we need to think of the mature phenotype being the outcome of a cascade of interactions between the environment and the organism which in turn depend on past interactions—at every stage that interaction is determined by the nature of the environment, the particularities of the genome, and the previous interactions between the environment and the organism.

When does life start?

This is one of the more politicized questions in science and medicine (we will leave aside the answer from the old vicar that it starts after the kids have left home and the dog is dead). It is a question given impetus by the various political and religious debates over abortion and the limits that might be placed on it. These are extremely important issues of personal choice, but they do not constitute a valid scientific question. Is it when the egg and sperm were each formed, or is it at fertilization, or is it at some arbitrary point in the chain of development from the early embryo to a newborn baby? Is it when the baby's organs are formed (and what does that mean given that there is progressive development of many organs such as the brain until well after birth)? Is it when there is the first muscular activity leading to limb movement, or when the first brain waves appear, or when the fetus starts to have a sleep pattern equivalent to dreaming (but what is it dreaming about?). Is it when the baby is capable of independent life if delivered (at least with the support of modern neonatology and technology)?

. . . and so on. These are not scientific questions, they are questions of personal and community values. Scientifically, life started about 3.8 billion years ago and since then there has been an unbroken chain of replication of DNA through to us and also to every other living animal on the planet.[2]

Even arguing that life starts at conception has its difficulties. True, our full genetic complement was not created until one of our father's sperm entered our mother's egg. But the way in which sperm are formed and the way eggs develop are very different. Post-pubertal males produce sperm continuously throughout their reproductive lives and store it from ejaculate to ejaculate in their seminal vesicles. So depending on the frequency of our father's sexual activity, the sperm that contributed 50 per cent of our genes in all probability was formed in his testes on the days before our conception. But in contrast, all the eggs a woman possesses are formed in the first few weeks of her intrauterine existence. So half the genetic material at conception may be only a few days old, but the other half will be many years old.

Why is male and female production of the sperm and eggs so different? Evolutionary biologists would reply that it is easy to see how across the animal kingdom there is advantage in each ejaculate containing lots of sperm so they have to be made continuously. If the female mates with more than one male around the time of ovulation, then it is more likely that the sperm that will win in the conceptional competition comes from the larger ejaculate.[3] It is like a lottery—the more tickets you buy, the greater your chance of winning. In fact there are lots of cunning ways males and females have evolved to try and deal with this sperm competition.[4]

The egg which contributes to the baby's genotype has already had a long life and may have been subject to a range of environmental factors, whilst the sperm has probably had a very brief existence before fertilization. Is it possible that the evolutionary advantage of the mother developing all her eggs before she is born derives from the opportunity this presents for them to be influenced by the environment inside the grandmother's womb—a sort of female memory effect? The egg from a woman who conceives in her forties is much older than if she had conceived in her twenties. Older eggs are probably not as robust as younger ones and may have been subject to the ravages of ageing. This may explain why women become less fertile after their mid-thirties. This is a biological reality that has become a social concern as more women choose to delay having a family until later in life, and then find that their fertility is not as they had hoped. The large increase in use of assisted reproductive technologies is directly related to this biology. There are evolutionary explanations for why fertility starts to fall after the age of 35, and we shall explain these later when we discuss the menopause.

From egg to body

After fertilization, the single-celled embryo with its full complement of genes—half from mother and half from father—starts to divide rapidly, each cell yielding two daughter cells with the same amounts of genomic DNA. These cells are embryonic stem cells—capable of differentiation into all the different cell types of the body by a series of further divisions, followed by highly specific phenotypic changes. Because these embryonic stem cells are able to turn into any cell type, they could theoretically be used to restock the cell populations of organs in which ageing and disease has killed some off, for example the brain of a patient suffering from dementia, the heart following a myocardial infarct, or the insulin-making cells of the pancreas in someone with diabetes. But sourcing stem cells for research purposes has become a matter for major current debate. There are theoretically enormous scientific advantages to the use of embryonic stem cells because these have the most potential to be manipulated into any cell type. In many tissues, such as the brain and bone marrow, some stem cells can be found throughout life but they are older, more limited in their potential, and, like an older egg in a woman's ovaries, may have suffered the ravages of ageing. More recently stem cells have been found to be present in the umbilical cord and in umbilical cord blood and they at least allow for the concept of banking so that cells can be matched to an individual, to be used later in their life if need be. However it is totally unknown whether such banked cells can be used therapeutically in the future—this is an area of active research but one made rather difficult by the political and religious context of studying stem cells. This is one of those impossible debates where individual belief systems prevent a societal consensus.

Tissue differentiation involves the orderly change in the profile of genes expressed in the cells so that they take on their particular characteristics. The science of studying this process of differentiation is developmental biology. And it has exploded in the last two decades owing to our ability to use the tools of molecular biology to identify when genes are turned on in development, in which cell, and how they relate to each other. Each cell's fate is determined by chemical signals emanating from neighbouring cells. These many individual signals are used to coordinate differentiation and further development. This is important because the cells cannot develop individually but must relate functionally to their neighbours; a liver cell is useless if it lies in the thyroid gland. From one cell division to the next, as some cells differentiate in one way or another, as they move (attracted or repelled by other cells), and as they secrete chemical signals into the milieu in which they lie, the information originally resident in one fertilized cell is used to develop a complex organ.

What is surprising is the discovery of how conservative evolution has been. Many of the same genes are used to regulate development in simple organisms like the roundworm, *C. elegans*, which has less than 1,000 cells in its body, and in the human who has over 100 trillion. But after all we do share about 25 per cent of our genes in common and we are very distant cousins, separated from each other in our evolutionary family tree about 600 million years ago.

Within each cell, its characteristics as a blood-forming cell or as a muscle cell or a skin cell are determined by which genes are turned on (or off) and so which gene regulators are activated. These are the same switches we described in the last chapter, utilizing epigenetic processes which permanently alter the activity of different genes in the cell. A pancreatic cell and a thyroid gland cell both have the same genes (we can use the word genotype to refer to a cell as well as to an organism) but in one the genes controlling the pathways that allow insulin to be made and secreted are turned on and the genes that control the making of thyroxine are turned off; in the other the pattern is reversed.

Thus the development of organs and systems proceeds in an orderly way. Given that the processes of development must involve the genes, the idea of a genetic 'programme' for development became fashionable.[5] Because the development of major parts of the body such as the limbs, wings, and body segments of insects could be shown to be orchestrated by gene expression, it was thought until recently that all development must be organized, coordinated, and regulated at this level and that it was essentially immune from external influence.

The highways of development

Driving from San Jose to San Francisco, a satellite navigation system will tell you to go up Interstate Highway 101. But traffic might be bad (it is California after all) and you might divert to Route 480 and then join Route 280 to arrive from a different direction in San Francisco. Or it might be raining heavily and the traffic slows right down, but eventually you get there. The rain and the traffic have acted as environmental factors that have changed the course of your journey, but the outcome is the same—you end up in San Francisco (although there may have been a cost to your blood pressure and equanimity). But if there has been an earthquake on the San Andreas fault which destroyed the freeway bridges you may never get to San Francisco at all—although hopefully you can get home again. In that case your journey has been totally disrupted by the environment.

So it is with development. Environmental factors might totally disrupt development leading to abnormalities, gross or subtle. We have recognized that external factors can disrupt development for about fifty years. Drugs such as thalidomide, excessive alcohol, infections such as rubella, and ionizing radiation can all cause birth defects. There is currently much concern about the many chemicals now found in our environment. Not infrequently we read in the media of a cluster of birth defects in families living near a chemical plant which is ascribed to the exposure of future parents to chemical agents. This type of exposure has also been a concern in soldiers returning from the Gulf wars. While it is difficult to be precise about how many defects should be ascribed to such exposures, there are enough clear-cut examples to be certain that this is a cause of birth defects. However the disruptive signal need not be so foreign as radiation or warfare. If a mother has extremely high levels of glucose in her bloodstream during pregnancy, because of uncontrolled diabetes, this can be a disruptive signal and cause defects in fetal heart development.

But in more usual circumstances, environmental factors do not disrupt the developmental 'programme' but rather they tune it—the journey still ends in San Francisco. Indeed it is these more subtle environmental adjustments to development that create an optimal match between the offspring and its environment. While the fundamental basis of development lies in the genetic information contained in the conceptus, it is that other also evolutionarily selected set of tools, the developmental plasticity toolkit, which allows the organism to adjust its development to the environment around it. The purpose of these adjustments is to improve the match. It is a matter of very different time scales—the genetically determined design assumes a gross match based on the environment that existed as the species evolved. The developmentally plastic processes allow matching with the environment existing around the time of the pregnancy and potentially, as we shall see, with the environment that the fetus anticipates inhabiting later in life.[6]

These modifying environmental signals can be present from around conception until well into the neonatal period. Before birth they are transmitted to the mammalian embryo or fetus via the mother and the placenta. In amphibia, birds, and reptiles, and even in insects, the egg can be influenced by environmental factors. In mammals the most important environmental factors to consider are nutrient and stress signals because, other than obvious physical environmental factors, the availability of food and the threat of predation are the most important environmental factors affecting an organism's survival.

The nutritional state can even have an influence from the very beginning of pregnancy. Women who are very thin or undernourished at the time they conceive are more likely to give birth to babies who are smaller or who are born prematurely. Women who are obese are more likely to have infants at risk of developing diabetes. Pregnancy is a time when the mother must supply large amounts of nutrients to the fetus and the state of her storage depots at the beginning of pregnancy can determine how well her fuel supply system will work for the remainder of pregnancy. Her nutritional state can affect the levels of the nutrients in the fluid in the tube leading from the ovary to the uterus, and that is the environment of the fertilized egg for the first week after conception. It can affect the concentration of nutrients in the inside of the uterus where the developing embryo stays for a few days before implanting and beginning to form a placenta, and once the placenta is formed what the mother eats and her metabolism will determine the levels of nutrients which cross the placenta to feed the fetus. After birth the mother's health will affect her milk production and many other components of her behaviour will influence her interaction with her infant, and thus how well it is fed and thrives. All these processes, across this long time period, allow the processes of developmental plasticity to optimize development with a view to matching the new individual to its current and future environment—as always the long-term objective is to maximize the chance of reproductive success.

For many animals risk of predation or competition from their own species or another is important. If the risks of predation or competition are high, stress is generated and the organism may need to have quite different strategies to survive. The acorn barnacle from the northern part of the Gulf of California usually has a feeding hole in the top of its shell. But there is a whelk that likes eating these barnacles, and this species has a sharp spike that it sticks through the hole in the barnacle's shell in order to feed on it. When the barnacle population senses that there are many whelks around, presumably from some chemical signal, the next generation of barnacles develops with a curved shell so that the feeding hole is at the side rather than the top. This usually prevents the whelk gaining access.[7] The barnacles survive, but the trade-off is that they grow more slowly and are less fit reproductively, so when the whelk population falls (partly because they can't get enough barnacles to eat) the next generation of barnacles returns to their normal form. There are many such examples of predator- and stress-induced changes in development.[8]

Less dramatic stress can also induce changes in mammalian offspring. When maternal stress hormone levels are high, they can affect the fetus and lead to changes in development. These include changes in the way the offspring's stress response system works, so that it may be more able to withstand living in a stressful environment. Recently, a collaboration between scientists in Edinburgh and New York has revealed that women who were pregnant in New York at the time of the 9/11 terrorist attack and who developed post-traumatic stress disorder have now given birth to children who have altered levels of the stress hormone cortisol.[9] The same scientists also have data suggesting that the period of sensitivity for changing stress hormone metabolism for life extends into childhood. They examined survivors of the Holocaust and found that the youngest have altered stress hormone metabolism some sixty years later.[10]

As the organism gets older, its capacity for changing its developmental trajectory diminishes. There are real costs to maintaining plasticity, so compromises must be made. Imagine building the prototype of a new car. At some point, early in the construction, it is too late for the designer to change his or her mind and say I want a six-wheeled instead of a four-wheeled vehicle or a diesel instead of a petrol engine. As the car is built it gets too late to change from having a manual to an automatic transmission. It even gets problematic at some point to change from leather to cloth seats, or to alter the paint from British Racing Green to Ferrari Red. The earlier changes cannot be reversed; the later ones can but the cost gets very high. A few choices can be left to the last minute: a CD player or a navigation system? So it is in human development—the later in development the harder it is to change the phenotype and at some point plasticity for many systems is essentially lost. For the car plasticity for axle number was lost very early, plasticity for the gearbox somewhat later, and for the paint and sound system later still. But plasticity for the tyres is never lost. Similarly in mammalian development, for some organs plasticity is lost early in fetal life—for example the primitive gonad is determined irreversibly as either an ovary or a testis at seven weeks after conception. For other organs such as the kidney, the number of filtering units is not fully established until late in gestation; and for the brain plasticity extends throughout life, but diminishes with age. There can be big species differences in plasticity. In contrast to a human who can only develop a limb in embryonic life, the mature axolotl can regrow a replacement if a limb is severed.[11]

Now or later?

There is now a growing recognition that plasticity is largely about establishing a better match between the organism and its environment—that is ensuring the organism will live in its comfort zone. If the environment has shifted then the organism must shift its phenotype so as to be in a new comfort zone. We can define two forms of developmental plasticity that are distinguished by *when* the adaptive advantage appears. If the environmental conditions in early development are severe the fetus may have to make some *immediate* adaptive responses just to survive. This would be equivalent to car engineers facing a financial crisis early in the construction stages; if they will not have the funds to finish the car according to the current plan, they will either have to modify their plans urgently or give up the project altogether. But most plastic responses in development are designed to tune the phenotype for the conditions expected in later life—we term this class of responses *predictive* adaptive responses.[12] This is equivalent to the car-makers being told the kind of conditions the car will be driven in while they are building it—it will be the same car, but its engine tuning, tyres, and climate control system may be changed.

The most common immediate adaptive responses are those of reducing early growth. This type of response is induced when the supply of nutrients from the mother to fetus is poor and the fetus must reduce its growth rate just to survive. Another kind of immediate response is to accelerate the maturation of the fetus so that it is born early. This will give an immediate advantage if the environment within the mother is so threatening that premature delivery might be a safer bet.[13] There is always a trade-off incurred with such immediate adaptive responses. Animals, including humans, born smaller or earlier are less likely to survive for a long time, but if they had not made the adaptive response they might not be alive at all.

But from the perspective of understanding match and mismatch in life after birth, our focus is on predictive responses. The embryo, fetus, or neonate uses them to try to adjust its biology such that its constitution will be better matched to the environment in which it predicts it will live, grow, and reproduce.[14] It does so by sensing its environment, using that information to predict its future environment, and then utilizing the processes of developmental plasticity so that its resulting phenotype will be better matched to that anticipated environment. The signal might be one component of nutrition, say a reduced level of an amino acid, glucose, or a vitamin, but the response must be integrated to allow the whole organism to adapt nutritionally, and survive and reproduce in an environment that it predicts will be inadequate. It must adapt many aspects of its biology and its whole life-course strategy to fit the predicted environment.

The better the match, the greater the chance of reproductive success. As a gross simplification the choice can be seen between interpreting the future world as short of food and high in competition and hence risky, or as having abundant food with little competition and therefore safe. The strategy to live in either world is very different. A risky world means not planning a long life, but maturing early so as to reproduce so that the individual's genes pass to the next generation. A safe world gives the individual the chance to grow larger, have more progeny, and potentially to be a bigger winner in the game.

After birth

The capacity for developmental plasticity does not end at birth although it becomes more limited as we grow. In many ways the neonate is as dependent as the fetus for information from its mother. But the human pattern of growth after birth is unique. We are born fatter than any other species,[15] and then have rapid growth in infancy and early childhood, slower growth through a prolonged childhood and juvenile phase, then we turn on our reproductive hormones, enter puberty, and go through another phase of rapid growth before finally our growth plates fuse and growth stops. So we can divide our postnatal growth into four phases.[16] The first is infancy, when we are initially totally dependent on breast feeding for nutrition, and which is characterized by a period of rapid growth which ends at about 3 years of age. In traditional societies weaning normally takes place at this time. A childhood period of slow growth follows, from 3 to 7 years of age, in which the child is still highly dependent on maternal care for protection and nutritional support. Then there is a juvenile stage with continued slow growth followed by a period of rapid growth again in adolescence.

Most mammals show rapid growth after birth or weaning followed by slower growth and sexual maturation. There is no obvious childhood period—once no longer an infant, most pre-reproductive mammals become juveniles which have little or no dependency on their mother for nutritional support. But no other species, not even our closest cousins the great apes, has an equivalent of our dramatic skeletal growth spurt at puberty although the male gorilla will put on much weight at puberty. It has been suggested from examination of fossil specimens that *Homo sapiens* and perhaps our most immediate ancestors are the first and only hominids to show such a dramatic skeletal growth spurt during puberty. The Turkana boy whose fossil skeleton is the finest example of the species *Homo erectus*, and who lived 1.6 million years ago in East Africa, reached a height of 160 cm. There remains debate as to how mature he was at the time when he died. He was clearly not fully mature but whether he was 7 or 11 years old is uncertain.[17] The problem is simple but the question remains unanswerable—the pattern of growth and maturation is different in every species and we will probably never know the pattern of maturation of *Homo erectus* and whether our pattern of a pubertal growth spurt is unique to *Homo sapiens* or appeared in any of our hominid ancestors.

So we have several questions about the unique pattern of childhood growth: why are we so fat at birth, why do we have a prolonged childhood phase not seen in other mammals, and why do we have a skeletal growth spurt at puberty? The answers seem to lie in the unique trade-offs we have had to make to be a large mammal which walks on two legs and which has a particularly large brain. Evolving to have a large brain necessitates that we are born in an immature state to avoid the problems of the narrow pelvic canal relative to the size of the head. We can see the results of the trade-off if we think about our neurological maturity at birth—whereas other primates have full locomotor ability at birth we are a year old before we can make our first tentative steps and 3 years old or more before we can run to keep up with our parents. This poses immediate problems for a hunter-gatherer family group on the move. Until about 3 years old the infant must be carried (and possibly suckled) in mother's or grandmother's arms. It is important to remember that interpreting human evolution must be done in terms of the environment in which the bulk of our evolution occurred—that is in East Africa as hunter-gatherers living in very small social groups.

Our degree of maturity at birth is more compatible with fast-reproducing smaller mammals. But we start a rapid period of brain growth before birth and this continues unabated in the months afterwards. This brain development must be protected at all costs and it is thought that the high-energy fat reserves of the human neonate serve the purpose of providing an energy buffer for this growth to continue even if nutrition is compromised.[18] We grow rapidly during infancy but then go through a very prolonged period of childhood when we are still not capable of fully independent living even though soon after our weaning our mother is likely to be pregnant again. Why do we have this prolonged period of childhood development? There are several theories; is it just a by-product of our long lifespan and generally slow tempo of maturation? Is it to allow continued growth of the brain and learning of life skills, or is it to allow a mother to wean earlier and rely on other members of the clan to support the infant during her next pregnancy?[19] Various anthropologists claim data to support each of these alternatives.

The juvenile phase is seen in all species of primate. Juveniles are no longer dependent on the mother for support but they are not yet sexually mature. Brain growth is largely complete but animals are learning the social and other skills required for their species. They must also play a supporting role in childcare in the colony—this is also true of humans in existing hunter-gatherer societies. It is important that this is before they enter the sexual competition. So once again choices about the timing of events and maturational phases in the life course have had to be made during evolution.

There are enormous differences in the longevity of different species and again these differences relate to the various strategies they have evolved to protect the continuity of their gene transmission. The wood mouse lives less than two years, the African elephant more than seventy. Some tortoises live very long lives—at the time of writing 'Harriet' in Queensland zoo had just turned 175. Most animals, but not human females, have the capacity to reproduce for most of their adult lives although in some of them reproductive performance declines with age—particularly in females of longer-lived species. There is debate about whether menopause is unique to humans, but we will return to this in a later chapter.

Answering the question of why we age is not easy. The most commonly held view has been that the mechanism of ageing represents the accumulated impact of wear and tear and exposure to reactive oxygen free radicals and environmental toxins which eventually lead to critical failure of some cellular functions. An evolutionary model is provided by reference to trade-off theory. Cellular repair and maintenance in the face of oxidative stress and environmental toxins is an energy-consuming process. But it is important to maintain cells in good order until reproduction is complete. So a trade-off is made—invest in cellular repair and maintenance, with its high energy costs, in early life and through reproduction (including parenthood) then reduce that investment.[20] Irrespective of how and why we age, natural selection will operate to select genes which favour successful reproduction and there will be little or no selection for anything which is beneficial to longevity once reproduction is over.[21] Thus selection could act to favour genes that give very high reproductive success even if they lead to death not long after. This is essentially what has happened in the male salmon—reproduce in a single episode then die. Thankfully it is not that simple in *Homo sapiens*.

Patterning our lives

Guppies are small tropical fish most often seen in home aquaria, but their real habitats are the streams of Venezuela and Trinidad.[22] But all the guppies in a stream do not look the same. Those living upstream have brighter iridescent spots and blotches including lots of blue while those living downstream are much less colourful, the spots are very small and the amount of blue is minuscule. The ones living upstream are also larger than those that live downstream. Why is this the case? The major threat to the guppy is from the larger carnivorous cichlid fish which live downstream. However, female guppies like large colourful mates so the upstream guppies have been selected for gaudiness and mate many times while growing to a large size before finally being eaten. But downstream there are a lot more predators. The guppy must be more wary and not advertise its presence. To survive the downstream guppy has been selected to be better camouflaged, and they stayed smaller and less obvious to the cichlid predators (although they still have enough spots to attract females of their own species). In other words, the life of the downstream guppy is more precarious and it has adopted a strategy that to pass its genes on to the next generation it must live in the fast lane. It doesn't invest too much in growth and advertising. The upstream guppies can be more leisurely and grow to a larger size, which in turn will give them an adaptive advantage. The processes that drive these different routes for development involve natural selection which has allowed different populations of guppies to adapt to different environments. It illustrates the point that the choice in early development of a life-history strategy which optimizes a match with the environment is crucial to survival.

Some environments show cyclical changes that are predictable, for example seasonal changes. In several species of butterfly or moth there are very different wing colorations depending on the time of the year in which the larvae hatch. For example the East African butterfly *Bicyclus anyana* can be born all year round. But those born in the wet or dry seasons look totally different—and for years they were thought to be two distinct species. Those born in the wet season have distinct bullseyes on their wings, but as the dry season comes, temperatures fall and food becomes scarce. The butterflies born in the dry season do not have bullseyes on their wings and are camouflaged to look like the brown leaves on the forest floor. These differences give a selective advantage because the bullseye-patterned wings distract predators from attacking their bodies—they attack the wings instead; this is less critical when food is plentiful, whereas in the dry season a different form of camouflage is necessary. The different appearances have been induced from the same genotype by environmental influences, in this case the ambient temperature and the length of daylight, which alter the release of hormones within the larva and control the expression of genes determining different wing colour patterns.[23]

Many mammals in the wild also have to take account of the seasons. Some of the clearest examples come from various species of voles. We have already described how the Pennsylvanian meadow vole adjusts its coat thickness to the seasons. For a related species, the mountain vole of the American Rockies, the optimal time to be born, to grow, and to mate is the spring and summer. This is when temperatures are highest and food supply is best, so that the voles can attain sufficient weight and reproduce early in summer, leaving themselves in good condition to survive the winter. But the young voles born later in the year will slow their maturation to conserve their fuel stores to allow themselves to tide over winter—if they had matured faster and reproduced at this time both they and their offspring would be unlikely to survive the winter. So the vole that lives in this niche environment has evolved a different strategy for dealing with seasonal variation. The change in developmental trajectory is determined by day length. Intriguingly, voles of the same species living in less extreme environments do not use this strategy. Like the newt, the developing vole has a genetically based but environmentally chosen set of alternative developmental pathways. The choice is made in response to environmental signals.[24]

Some other animals use an even more remarkable delaying tactic—they can push the 'pause button' on pregnancy—a process known as embryonic diapause.[25] Rather than implanting, an embryo can remain in suspended development floating in the womb. This process enables the female wallaby to be a continuous production line for young joeys. As soon as she gives birth, she immediately becomes pregnant again. But baby wallabies are born and attach to the pouch in an extremely immature form and all her energy supplies are put into supporting the infant on the nipple. To help do this, embryonic development is suspended, but as soon as the joey in the pouch is weaned embryonic development is restarted and the next baby wallaby is born soon after (as pregnancy is only twenty-seven days). Again she conceives and that embryo is put into suspended animation until its elder sibling is weaned, and so on. If the joey in the pouch is lost, the quiescent early embryo is reactivated, implants, and resumes development. Using delayed implantation is also a very successful strategy in the roe deer. The doe conceives in summer when she is in peak body condition and thus most fertile. But by delaying implantation she does not use up excess nutritional reserves during winter. In the spring the pregnancy continues, allowing her to support the growth of twin or triplet fetuses when food supplies are again plentiful. A wide range of other species have such seasonal diapause including bats, armadillos, and skunks.

Climatic and nutritional constraints on growth and development are of paramount importance to many species, and we can see the survival advantage of being able to adapt to them. As we all know, it is not only important to be able to cope with the weather but to be able to forecast it. It is awkward to be out on a day that turns out to be hot when you have brought an overcoat, a scarf, and an umbrella. It is even worse, and downright dangerous, to be caught wearing shorts and a T shirt in a blizzard on a mountain when a day which looked fine at the start has turned out otherwise. The key thing about a forecast is that it is made in advance, to allow preparation for the event before it happens. For life-course strategy, the more accurate the forecast, the higher the chance of survival and hence of successful mating.

This predictive process begins before birth, enabling the developing embryo and fetus to make adaptive changes before it is born, in order to be better able to survive in the predicted environment. These forecasts must use cues from the mother that cross the placenta to the developing organism and inform it about the world outside. Some cues can also be transmitted to the offspring while it is being suckled, because the infant does not face its nutritional environment unaided until after it has been weaned. But the key point is that the sooner the mother can inform her offspring about the world in which it is likely to live, the more accurate will be its tuning of its plastic response and the better will be its chance of survival in that world. So offspring use the information transmitted from the mother during development to predict the nature of the future environment, and to alter the development of tissues, organs, and control systems accordingly. When the prediction is correct, the offspring will have a survival advantage. But if the prediction is inaccurate the offspring will be at risk by being mismatched to the environment it will face. It turns out that even though the prediction can be wrong and generate mismatch for some offspring, provided that the prediction is more often right than wrong it will often be of evolutionary advantage.[26] For many of us living on islands like Great Britain or New Zealand, we would be delighted if the weather forecast was right more than 50 per cent of the time.

We are now learning a good deal about the processes by which offspring are able to make these predictive responses. The signal from the mother may be a change in the level of a key nutrient, or of a hormone in the bloodstream. For example, both in the mountain vole and in another related species, the Pennsylvanian meadow vole[27] in which the pups are born with different coat thickness appropriate for the impending winter or summer, it is the changing day length during pregnancy which is the cue. The fetus senses this because the hormone melatonin made by the mother can cross the placenta. Melatonin levels cycle between low levels during the day and high levels at night and are set by the length of the dark period. So when the days lengthen in spring, the shorter periods of high melatonin concentrations in the mother's bloodstream tell her fetuses that summer is coming, and in the autumn, when the days shorten, the higher levels of melatonin inform about impending winter. The voles' development is changed accordingly.

During development, several aspects of the environment are signalled by the mother to her fetus. They include levels of food availability, the numbers of predators around, and social conditions. Animal studies show that a period of fluid restriction during pregnancy, to simulate drought conditions, alters the set point of the offspring's salt and water regulating mechanisms to make it more able to conserve water. All these give the offspring an adaptive advantage in their future environment if the prediction made during development is appropriate.

But the prediction can be wrong. The environment may have changed between embryonic or early fetal life and postnatal life. And if the mother is sick, if she is on an inappropriate diet, or if the placenta is dysfunctional then the fetus might predict undernutrition in its future when in fact that will not be the case. If the mother smokes, nutrient transport across the placenta is inhibited by the action of nicotine[28] and this leads the fetus to make the prediction that it will live in a world of poor nutrition, and to adjust its phenotype accordingly, when in fact there is no nutritional limitation. We shall return to this problem in Chapter 7.

Although we have used examples of illness to make the point, the reality is that all fetuses are making predictions about their future all the time. They all try to optimize the match between their constitution and the environment they will face. So anything that can affect fetal development can affect the later match. The entire life-course strategy of the organism is tuned by those early experiences.

The right road

Comparative biologists usually think of life strategies in terms of two major factors: food and sex (most teenagers are not so different). Different life-course strategies represent the result of selection, based on the degree of the match between the biology of organisms and their physical environment, the availability of food, and the threats that exist within the environment (physical, predation, or competition from other members of the same species). There are basically two extreme classes of strategy that can be adopted, and while some animals use these extremes, the strategy adopted by most species lies somewhere between these.[29] At one extreme are animals that reproduce in enormous numbers but in which very few of the offspring will live to reproduce. These animals normally have very rapid development, they may have a very short lifespan as a mature adult, and they have small body size. Insects, fish, amphibia, and many small mammals such as mice follow this general pattern, albeit to different degrees. At the other extreme are those animals which have very few progeny but invest highly in these progeny. They show a much lower infant mortality, they grow slowly to reach a relatively large body size, and they reproduce later in the life cycle. Examples include the elephant, the horse, the blue whale, and the human.

Now these basic strategies can be modified by environmental factors. For example in a threatening environment, the appropriate response to a poor nutritional environment for a high-volume, short-lived breeder may be to accelerate maturation and reproduce early; even if that puts the individual organism at risk, gene transmission has been preserved. In contrast in the slowly maturing, later reproducer which has only one offspring per pregnancy, and in which maternal survival is necessary to support the juvenile through its long postnatal development period, reproduction will be delayed in similar circumstances; the animal must delay reproduction in the hope that times will become more propitious. Most animals have a life-course strategy somewhere between these extremes but humans are an example of a very slow developer and our life history can best be understood in these terms.

Just as in other animals, humans evolved with the capacity to make biological trade-offs in development. We show interactions between nutrition, development, growth phases, longevity, and reproduction.[30] When a human fetus is undernourished *in utero* it trades off growth for earlier maturation and a premature birth. In many species increased reproductive success leads to shorter life in the female, another form of biological trade-off. Studies of longevity in the British aristocracy (who are presumed to have been living in optimal socioeconomic conditions for their times and for whom the reproductive records are relatively reliable) reveal that the more children the queens and princesses had the shorter were their lives.[31]

Humans are one of over twenty species of hominid that derived from a common ancestor who evolved to walk on two legs several million years ago. We have some unique characteristics as a species and these define our pattern of development. We have a big and powerful brain inside a large skull—our brain size (1,350 cc) is almost twice that of our early ancestor *Homo habilis* (800 cc) and four times that of the first hominids. Walking on two legs necessitates a developmental compromise if the baby is to be delivered through the pelvic canal—which, until the relatively recent development of caesarian section, was the only way a live baby could be born.[32] To be efficient on two feet requires a change in the positioning of the hips and a narrowing of the pelvic canal—otherwise we would fall on our faces every time we tried to walk. When we compare the human baby's head size at term to the width of the pelvic canal, there is not much room for error. This contrasts with most other primates where the pelvic canal can be bigger because they do not walk upright. Thus while a chimpanzee baby can pass straight through the pelvic canal face forward, the human baby cannot. The only way it can be delivered is to rotate its head to get past the narrowed angles created by the repositioning of the hips. Consequently the human baby is unique in not being born face forward. The anthropologist Wenda Trevathan[33] has suggested that the human infant's chances of survival would be greater if there was a second person present to clear the baby's mouth so that it can gasp for air as its head appears. Was this the evolutionary origin of the midwife?

Nesting mammals such as rabbits, stoats, and field mice have short pregnancies, and give birth to a large litter of very immature infants who need the safety of the nest to continue their development. They cannot see, their skin is very fragile, and they cannot walk or control their body temperature well. But their postnatal maturation happens fast. A rat pup may not be weaned until twenty-one days of age but it is through puberty and able to mate only thirty to forty days later. The technical term for this type of species is 'altricial' and in general these species are small with short lifespans. Marsupials are a line of mammal for which the ancestors diverged from those of other mammals some 150 million years ago. Their pattern of development is even more extreme. Their offspring are essentially embryos when they are born. They follow a trail of chemical scent to the nipple where they attach, then undertake an enormous amount of their maturation there rather than in the uterus. Monotremes (the name means 'single opening' in Greek and refers to the reptile-like use of a common duct, called the cloaca, for urination, defecation, and reproduction) such as the platypus represent an even more ancient branch of the mammals, now confined to Australia and New Guinea. Reproduction in monotremes is unique amongst mammals—they lay eggs rather than having any intrauterine development but still suckle their young after hatching through primitive milk glands that are no more than specialized patches of skin.[34]

Other animals, particularly the larger mammals such as the hippopotamus, horse, and moose, give birth to very mature offspring which, while dependent on their mother for food, can move around, have good vision, hearing, and muscular control, and regulate their body temperatures well. All the apes are like this except the human. To solve the problem of the large fetal head but narrow maternal pelvis we evolved to give birth to a very immature baby which needed to grow its brain substantially after birth. So unlike the other apes we are born with a rather immature brain, unable to move, unable to seek out our mother, and totally dependent on her. This determines the long period of infant and childhood support we require until we are fully independent and it has influenced the type of social structure humans evolved to allow a prolonged period of parental (usually maternal) support.

So having big brains and walking on two feet have had their consequences. Our pattern of development has been determined by the need to strike a balance between these two competing demands and in this way human development is no different from that of other species which also have to make trade-offs as a consequence of the life-course strategies they adopt. The human fetus receives cues about its environment and adjusts its development to tune its life-course strategy. It tries to predict its future environment. We are not infinitely plastic and this limits the environments we are designed to inhabit. But we have one other unique attribute: we are highly skilled at modifying our environments—we turn our attention to this aspect of our biology in the next chapter.

4
Things Ain't What They Used to Be

Much of northern Nigeria has been riven by wars and repeated famines. The further north we travel, the more arid the country becomes. The far north-east, on the border with Niger, is mainly acacia scrub in which it is just possible to scratch an existence by cultivating millet and keeping a few bony cattle. The summers are hot and dry, and although the rainy season brings relief it also turns the crumbly soil into a slurry into which the wheels of even a bicycle instantly sink up to the axle. This region is cut off from trade, communication, and from medical supplies. It seems to be a place so far removed from modern life that visiting it is like taking an enormous step back in time. Indeed this region of Africa may well be the cradle of humanity; recently the remains[1] of the oldest known hominid, an individual nicknamed Toumai who lived about 7 million years ago and had a brain size about 25 per cent of that of a modern human, were found there. Yet it would be wrong to say we would be stepping back into prehistory. Even for people living there today life is probably very different from what it was in Palaeolithic times when *Homo sapiens* first evolved.

Many of the area's tiny settlements of Kanuri tribespeople lie close to Lake Chad. This is a veritable inland sea albeit considerably smaller than it was in the past. A few other villages are strung out along the rivers that feed into the lake. Beyond this, the area is virtually uninhabited, because the problem of water—so little in the dry, and far too much in the wet season—makes human life impracticable. With the help of international organizations and a large investment of US dollars in the late 1960s, the solution to obtaining a better life and prosperity in the region presented itself. A plan was hatched to divert the waters of the rivers flowing to the lake into a system of irrigation canals, permitting a continuous supply of water to new settlements and the cultivation of valuable crops such as rice. It was predicted that the economy would grow, making the construction of sealed roads and even an airport feasible. Some agencies even envisaged that Lake Chad could eventually become a tourist resort.

Unfortunately this solution, like many other attempts by humans to manipulate their environment, was based on a faulty premise. It ignored one of the main reasons why humans have always had a marginal existence in this region. This is the tropical disease bilharzia, and it was this disease which one of us (Mark) travelled to Nigeria in 1971 on a medical research expedition to study. The disease is chronic and endemic and affects over 170 million people in sub-Saharan Africa. In rural Nigeria alone over 70 per cent of children and 50 per cent of adults are affected.[2] It is caused by a tiny parasite that infests the liver and spleen, causing enlargement and slow failure of these organs. Like many other human parasites it has a complex life cycle and does not pass directly from one human to another. The parasite passes its eggs from its human host through their urine and faeces. Therefore where there is poor sanitation the eggs get into the rivers and ponds of the area, especially in the rainy season. The larvae that emerge from these eggs[3] then infest a secondary host, this time a water snail, from which they emerge as a free-living form, a minute worm-like creature which can swim to find a new human victim. This might be a child bathing in the river or a woman washing clothes on stones at its bank. This form of the parasite can penetrate the intact human skin to get into the bloodstream, and from there it will take up residence in the liver and spleen and so the cycle continues. The disease spreads very effectively from human host to human host by this strategy during the wet season. But the snail has been a clever choice of secondary host in which the larvae can incubate, because snails can survive drought for many months buried in soil, only to become active, breed, and release the skin-piercing worms once the rains arrive.

Humans will have lived with a level of infestation by the bilharzia parasite for millennia as they drew water and washed in the rivers, or fished in Lake Chad. This is part of the cost of inhabiting this particular area of the globe. The developers of the irrigation schemes, which were still in a pilot phase in the early 1970s, knew that the canals they wanted to develop would soon be populated by water snails and that humans cultivating cash crops such as rice in the new paddy fields would be exposed to the skin-piercing worms. But so what? Their argument was that because a high proportion of the Kanuri farmers already had the disease, further exposure to water would make little difference to its incidence. Mark's studies in Oxford had brought him into contact with tropical disease experts who thought otherwise. The severity of the symptoms is directly related to the parasite load. As the parasite cannot pass from one human to another without first infesting the snail, the degree of ill health depends on a person's exposure to water. The Kanuri farmers or fishermen might already have the disease, but to an extent their society had become adapted to this unavoidable cost of living in this environment and their parasite loads were not usually so high that they could not function. But once they started to work on the irrigation scheme they spent more time in the water and the level of infestation got substantially worse.

This sudden change in their environmental exposure and parasite burden affected the Kanuri badly and the consequences were devastating; their workload and productivity fell, previously productive men were no longer able to work, and household income fell—with both nutritional and social consequences. From being a community in a stable subsistence state, a sudden change in their ecosystem would put them in a situation dependent increasingly on external aid. Here was a community in which its members had adapted, albeit at a cost, to a particular and marginal environment. When their own species interfered with and further changed that environment, it placed them in a situation beyond their inherent capacity to cope and the consequences were severe ill health and social disruption. The Kanuri were in trouble because they, or rather other humans who thought they knew what was good for them, had changed their environment. Here was a paradox: humans have been a successful generalist species because we can manipulate our environments to extend our comfort zone; but sometimes that very manipulation has detrimental effects on ourselves.

Manipulating the environment

Humans are not alone in manipulating and controlling their environment. Many insects such as termites and hive-building wasps control their environments by building them. The termites of Australia's Northern Territory build mounds several metres high.[4] These have a flat blade shape and are oriented north–south. Many travellers have been puzzled by these so-called 'magnetic' termite mounds, but are also grateful for their navigational help in a vast landscape. The termites have adopted this north–south design to maximize stability of the temperature inside the mound. The flat sides absorb as much of the sun's heat as possible in the early and later parts of the day, whilst the effects of the intense midday sun are minimized because at that time the edge of the mound faces the sun.

Most species do not build their own environment, and specialist species are most successful within the very closely defined natural environment where they evolved and to which they have become uniquely adapted. If they cannot evolve or adapt to an environmental change, they may be at risk because their adaptations have left them with a very precise and highly specialized comfort zone. Often these ecological niches are very specific—for example, the thirteen surviving species of finch on the Galapagos have each adapted to their specific niche defined by the plants they prefer and the seeds they eat and whether they prefer to be on the ground. These are perhaps the most famous bird species in the study of evolutionary biology as they have been extensively studied by the Grants and demonstrate various aspects of evolution in action.[5] Although specimens were collected by Darwin he never studied them in detail, and contrary to popular mythology they were almost certainly not the critical components of his early thinking—he only made a passing reference to them in *The Origin of Species*, and indeed in a problem feared by every scientist he did not label clearly which birds came from which island and from which niche.[6] Some of these birds live on the ground and rarely fly, others live higher off the ground. These various finches are distinguished by different body size and beak shape and size and their preferred food sources. It is the characteristics of their beaks that determine which seeds they can eat. These many species evolved from a single precursor species to fill all the various niches on the islands and thus avoid too much direct competition.[7] But when times get tough, for example after a prolonged drought, some types of seed disappear—species with beak shapes that do not match the available seeds do poorly and within a species those with the least suitable beak shape do not survive. Thus the weather shifts both the mix of species and the characteristics of beak shape within a species. Weather is generally cyclical, and so over the years the balance of characteristics cycles with the weather. But if conditions shift too far or shift irreversibly in one direction or another, there is a high probability that a species will be lost. Indeed, one species of finch that Darwin collected (and it was the only specimen of that species ever collected), the great Galapagos finch of Floreana Island, *Geospiza magnirostis magnirostis*, was extinct only a few years later. It is probable that the reason for the extinction was the extinction of their major source of food, although not because of the weather. These birds almost certainly lived on a local cactus with the biggest seeds of any Galapagos cactus—their specialized niche was to be the only bird able to feast on these large and hard seeds and their beaks had evolved accordingly. The cactus became extinct because cattle were left running wild on the island after a prison colony was abandoned. The finches were also largely ground birds and would have made a tasty meal for the prisoners' cats, so their environment was affected in a second way. So here is an example of the extinction of one species leading to the extinction of another specialized species that could not adapt to cope with the environmental shift.

But whereas 'niche constructors'[8] such as the termite build their own relatively constant environment and specialist species are restricted to very specific environments, humans are constantly creating new environments and inventing new technologies to live within these environments. Technology can allow a broad range of environments to become comfort zones. Clothing, igloos, and hunting tools allowed the Inuit to live in sustainable communities in the Arctic, and floating reed islands allow the Aymara people to live on Lake Titicaca in the Andes. Our technology enables us to construct massive skyscrapers and underground shopping malls, where we can live all year round at a constant temperature.

There appears to be no limit on how we can use technology to manipulate our environment. Indeed the more recent history of our species is full of examples of the use of new technologies, often to produce some very extensive modifications of the environment, whether physical, nutritional, or social. While this allows us to survive in environments beyond those we first evolved to inhabit, increasingly the environmental changes we create bring cost and that cost involves greater mismatch between our constitution and the environment, and therefore an increased risk of disease.

But first we need to look back at how human environments have changed during our history as a species and how humans have themselves created much of this environmental change. In a simplistic way the early part of human history can be seen as the *expansion* of humans across the planet as they learnt how to adapt to different environments, and our more recent history (roughly the last 10,000 years) as a period in which humans substantially *manipulated* their nutritional, social, and physical environments through the development of agriculture, settlement, and complex societies. It is the extent to which we can or cannot adapt to these manipulated environments which interests us. To discuss this idea, we will not try to describe human history in one chapter—we are not historians and there are many fine books on the subject[9]—but as biologists we cannot talk about match and mismatch without identifying those environmental factors which have changed since our species evolved, because it is the issue of whether we have the capacity to cope with these environmental changes which determines how well we are matched to our modern world. Thus the remainder of this chapter will focus on those elements of the environment which have changed—and in some cases have changed beyond our capacity to adapt and thus have generated cost. The focus will be on our unique capacity as a species to generate technological change, manifest as altered food supplies, longer lifespan, and living in much denser aggregations in complex societies.

On the move

It is generally believed that the first hominids moved from a forest environment to the savannah and that it was this transition which directly or indirectly led to the adoption of a crouching, then an upright, posture.[10] The capacity to make tools and to use fire appeared in *Homo erectus* on the eastern side of the Great Rift Valley in East Africa between 1 and 2 million years ago and the earliest member of *Homo sapiens* appeared in that region about 150,000 years ago.

Through this period there were major changes in the environment, and these can be studied from ice-core samples for the measurement of characteristics such as trapped gases. There were complex cyclical changes in the planet's mean temperature and water levels which were driven in part by changes in the shape of the Earth's orbit around the sun (about a 100,000-year cycle), in the Earth's tilt (a 42,000-year cycle), and a wobble about the Earth's axis (a 22,000-year cycle). These various cycles interact, and together with changes in the sunspot cycle and factors such as continental drift changed the pattern of ocean currents to generate substantial variations in global climate. Cooler periods were associated with glaciation, progressing from the poles, and a fall in sea levels as water was trapped in the polar ice—

these were the Ice Ages. The planet's history has been one of cyclic warming and cooling, albeit in very irregular cycles and changing patterns because of the range of factors involved. The most recent Ice Age began about 100,000 years ago and ended about 11,500 years ago; since then the planet's climate has been in a warm interglacial phase.

There were at least two important hominid migrations out of Africa. The first involved *Homo erectus* about a million years ago. That hominid and some of its descendants such as *Homo neanderthalensis* came to occupy a broad range of Eurasia—from Spain to Indonesia. Did they migrate as their population expanded or because the conditions were changing and so they were pressured into moving to stay within their comfort zone? We do not know. The Neanderthal and perhaps the dwarf hominid, *Homo floresiensis*, possibly a distinct species recently discovered on Flores Island in Indonesia, were the last two species of hominid to share the planet with our own species. Until recently it was debated whether *Homo sapiens* arose solely in Africa or evolved from *Homo erectus* and its successor species in multiple sites in Africa, Asia, and Europe, but now genetic marker studies indicate that *Homo sapiens* evolved only in Africa somewhere around 150,000 years ago.

The path of the migration of *Homo sapiens* out of Africa has been mapped by magnificent detective work based on two approaches—one that allows researchers to follow female migration and another male migration. A special form of DNA contained in small organelles within cells called mitochondria is passed only from mother to daughter (because there are no mitochondria in sperm). Conversely, the Y chromosome is passed only from father to son.[11] Knowing this gives researchers the tools to track both male and female lineages by looking for markers and mutations in mitochondrial DNA in females and in the Y chromosome in males. For example, this technology has been used to identify and track a particular Y chromosome pattern originating in Mongolia about 1,000 years ago which occurs in about 8 per cent of men at present living within the boundaries of the historical Mongol empire. The most likely origin of this lineage is suggested to be Genghis Khan himself—who had very many children and whose slaughter of conquered peoples caused 'selection' favouring his lineage.[12]

Homo sapiens started to migrate out of Africa about 65,000 years ago. By 45,000 years ago we had reached Australia; even then, in the Ice Age, this required a significant sea crossing from what is now Indonesia, and this suggests that human technology had evolved quite substantially by that period. It is probable that humans only entered the Americas some 13,000 years ago when the ice had retreated to allow a conduit across the Bering Strait and through an ice-free passage east of the Northern Rockies.[13] The last significant land mass to be settled by humans was New Zealand, where the Maori arrived by ocean canoe only 1,000 years ago, completing the great Australanesian migration which had started in Taiwan about 4,000 years ago.

The size of the migrating groups was probably quite small, maybe ten to fifty members who were related to each other. It is estimated that only a handful of individuals crossed the Bering Strait and their descendants include all the North and South American indigenous peoples. Similarly, a very small number of individuals appear to be the ancestors to the whole Finnish founding population.[14]

Throughout history nomadic bands have been able to survive by moving considerable distances. It may have taken only 1,000 years for the whole of the Americas to be populated from when the first nomads crossed the Bering Strait. As hunter-gatherer populations dispersed across wider geographical areas the environments to which they had to adapt became more disparate. Their technologies of fire, tools, clothing, building shelters, and communicative skills essential for group hunting allowed them to do so.

The Palaeolithic human

Language has given our species the capacity to interact, to plan, and to use technology and many experts think representational art and abstract thinking are not possible without the evolution of language. But we do not know when language appeared. At the time we started leaving Africa about 65,000 years ago, there seems to have been rapid development in our technological and sociological capacities, and one view is that this was linked to the development of our mature language capacity.[15] But others have argued that language must have developed much earlier.[16] Whichever is correct, sophisticated and specialized tool use was accompanied, or soon followed, by representational art on cave walls, the appearance of beads and other personal decorations and carved objects. Burial of bodies together with material objects appeared about the same time and this has been interpreted as evidence that by then our ancestors had some beliefs such as in an afterlife and traditions which might be considered 'religious'.[17] We can see the beginnings of a human culture that can fuel our imaginations. Was life now sufficiently secure that there was time for art? Or were the long winters in caves and the prospect of hunting in the spring, with the worry about attack from neighbouring groups, the driving force for such art? What were the hopes and fears of these people? Did they understand that life could be longer, and so they buried their dead with ritual to propitiate some deity, or was the burial a celebration of life? We will never know, but this does not prevent us from recognizing these as important signs of a crucial aspect of human evolution—the appearance of complex and abstract thought processes.

This period of human development is termed the Palaeolithic and our understanding of it is somewhat speculative, to say the least—archaeology and palaeontology have their limitations. The Palaeolithic era ends with the development of agriculture and settlement, which occurred in various regions of the world between 10,000 years ago and the present (some groups such as the Australian Aborigine did not adopt post-Palaeolithic technologies until recent times). Perhaps 95 per cent of our existence as a species was in the Palaeolithic. However it would be wrong to assume that human existence in the Palaeolithic was uniform—the range of environments we occupied was broad and there were major changes, at least in tool-making techniques and the use of art in the later Palaeolithic.

While our knowledge of Palaeolithic times is restricted to archaeological study of the limited material available, some researchers have tried to gain insights from studies of the few hunter-gatherer societies which remain to this day. While societies such as those Australian and some Papuan tribes which did not develop agriculture until European contact lived in a pre-Neolithic state, they were poorly documented (if at all) before being suppressed or exposed to modern technologies. There are still a few existing hunter-gatherer societies, e.g. the !Kung in the Kalahari desert and some Papuan and Amazonian tribes. But these have been squeezed to life on the very margins of the developing world and are often displaced from their preferred ancestral environmental range. They have been compressed into ecosystems they would not necessarily have inhabited and thus compelled to live in ways that have become very different from those of their ancestors. So extrapolation from observations on such peoples back to prehistory is very difficult.

The range of environments to which Palaeolithic humans were exposed was very broad, from tropical Africa to the frozen Arctic. But it is possible to make some generalizations which help us to understand the selection pressures operating on our ancestors—for example early humans evolved without exposure to large groups of people with complex social hierarchies but were equipped primarily to deal with small social groups. And they would have had no exposure to the higher-density carbohydrate and fat diets associated with agriculture. So there would have been no evolutionary pressures to deal with such exposures.

One particular concept, that of the 'environment of evolutionary adaptedness', was first introduced by the psychoanalyst John Bowlby and further developed by two evolutionary psychologists, Cosmides and Tooby,[18] to define the selective environment of the Palaeolithic, with particular reference to psychological function. They hypothesized that brain function evolved in a modular manner and that these modules would have been shaped by the social environment of Palaeolithic times—that is, small family groups living in relative isolation. However perhaps in making a valuable point that selection is environmentally determined they went too far. A key feature of the human brain is that it is self-learning and plastic, and it does not operate like a closed system. Thus, while we will have evolved with constraints on the less plastic components of our biology such as metabolic control, patterns of growth, development, and reproduction, this may not apply to our brains. Not all aspects of human behaviour can be interpreted purely as responses to a Stone Age environment. We prefer the view[19] that this is an overstated and oversimplified concept. But it does make one point—that to understand modern humans we must consider how we evolved and with what constraints.

Food and farming

Our ancestors in Africa lived as hunter-gatherers, relying on food from two major sources—the collection of seeds, tubers, nuts, and fruits, and on hunting. There have been various estimates of the content of this Palaeolithic diet and it is clear that it differed significantly in a number of respects from a modern diet. The Palaeolithic diet was higher in fibre content and had a much lower glycaemic index (ability to raise blood sugar rapidly) because the foods were less refined. Wild honey would be the only source of concentrated sugars and it would have been a minor component of the diet. The diet had a very different mix of fatty acids, a higher protein content, and a much lower salt but higher potassium content. There was no milk, butter or cheese, and the meat was generally much leaner than today.[20] Hunter-gatherers dispersed in accordance with available food supplies. They could choose their environments, within limits. There is no fossil evidence to suggest that they suffered chronic malnutrition—quite the reverse, because the data available from skeletons suggest that they achieved modern, or close to modern, heights.[21]

The last Ice Age ended about 11,500 years ago with dramatic changes in the landscape as sea levels rose and in vegetation when both temperate zones expanded. Temperate and tropical forests expanded. For human population groups this marked the start of what has been termed *intensification*.[22] Rather than just taking what they could from their environment then moving on, humans started to use technologies to extract more and more from a static environment. Domestication of animals and agriculture developed in different ways at several times in various parts of the world. While agriculture appeared about 11,000 years ago in the fertile crescent extending from the Levant to the Tigris and Euphrates, it did not develop in African regions until 4,000–6,000 years ago and never developed in Australia (until much later contact with European colonizers).[23]

As Jared Diamond has described in *Guns, Germs and Steel*, biogeographical factors, and in particular the nature of the local vegetation and larger species of wildlife, determined the particular path of development agriculture took in each region. The climate changes at the end of the Ice Age induced huge vegetation changes with corresponding major changes in the distribution of animals; as forests expanded the large herbivores of the Eurasian steppe were replaced by smaller species, making hunting more difficult and favouring the development of pastoralism. In the fertile crescent the shifts in climate and vegetation made foraging more difficult and favoured the development of agriculture from about 11,000 years ago.[24]

Both plants and animals were domesticated by selection, in this case by conscious or unconscious artificial selection to develop herds of animals and crop plants. Herded animals were generally smaller than their wild counterparts and plants were selected for early germination and larger seeds with thinner skins. Interestingly the experiments of Belyaev already discussed, in which Siberian silver foxes were bred for tameness, suggested that such selection exposes latent genetic and phenotypic variations which may have formed an important component of the development of agriculture.

What caused Palaeolithic societies to turn to agriculture? It is important to note that this switch occurred independently at least nine times in different parts of the world and then spread out from these distinct nodes. It is also clear that all societies did not necessarily turn to agriculture even when there was local knowledge of it—foragers and farmers coexisted throughout Asia. Presumably the combination of population expansion together with climate change was the major impetus for the development of agriculture. Some of the more affluent foraging communities had showed some form of semi-permanent settlement in Palaeolithic times in localities where there was sufficient food for them not to have to move constantly—for example the mammoth hunters of the Ukraine had semi-permanent villages with storage sites for meat some 20,000 years ago.[25] Technology development such as better fish hooks, spear-heads, blade edges for harvesting wild grains, and animal traps allowed for more effective hunting and foraging, thus supporting population intensification and sedentary lifestyle. This would have driven an increase in material exchanges between groups—the beginning of trading—indeed it has been suggested that the first towns appeared at exchange sites such as Jericho.[26]

Agriculture also brought a progressive change in diet—as plants became more selected for cultivation, herding allowed the collection of milk as a food source and there was access to fatter meat on a more consistent basis. The inclusion of milk in the diet is highly significant. To absorb cow's milk we need to have the enzyme lactase in the gut, to digest the sugar lactose found in milk. But ancestral humans are not thought to have commonly had this gut enzyme. So they were not matched to such a diet, at least initially. Now we find intestinal lactase throughout the European population, although less so in other populations. It has been suggested that humans who had lactase were positively selected in those farming populations herding cows, sheep, and goats and consuming milk, butter, and cheese. Paradoxically we have now come to regard lactase deficiency as an abnormal condition.[27]

But there were disadvantages to the development of farming. Herding and domestication brought humans into much closer contact with both animals and rodent pests and so there was an increased risk of infectious disease. Many viral diseases such as the influenzas and some parasitic and bacterial diseases such as leptospirosis and salmonella have their origin in domestic animal hosts. Often these viruses are endemic in animal and bird populations, where they may produce no symptoms. But the pandemics of flu in 1918, 1957, and 1968 show how devastating they can be once they have found a way of infecting humans and evading our immune defences. The pandemic of 1918 is estimated to have killed between 20 and 40 million people worldwide, far more than the First World War it so closely followed. The recent cases of avian flu in humans, caused by the H5N1 strain of the virus, first arose in China where humans live in close contact with many species of domesticated and wild birds, and they show how important this threat remains. They are the most recent in a long line of such infections reflecting the constant battle between the influenza virus and humans for reproductive fitness. And there is another side to this story too. Because we became adapted to living with high levels of some forms of bacterial pathogens, it is now suggested that our immune systems suffer from the lack of such threats in our modern, cleaner world. This may partly explain the increasing incidence of asthma and allergies, especially in children.[28]

Villages and towns

The development of agriculture brought very drastic changes in social organization. Villages appeared, followed by towns and then cities. Some people now lived in contact with much wider networks of others than they would have had if they had maintained the hunter-gatherer or pastoralist way of life. Echoes of the latter persist today. Much of the so-called developing world persisted with a hunter-gatherer/pastoralist approach until colonization. The strength of colonizing powers was based on their technological advantage arising from denser populations, networked collective learning, and highly differentiated skills. It enabled them to achieve technological dominance over people they generally viewed as more 'primitive'.

We will first discuss those peoples for whom settlement became the dominant way of life. It necessitated the construction of more permanent dwellings. It also led to groups of people living together, and as these groups got larger, skills started to be separated within society. Not everyone needed to do everything—perhaps some wove and some made tools and some tended the crops. Populations grew rapidly when times were good but settlement made it harder to move in years when the harvest was poor.

It is often thought that the life of the hunter-gatherer was one of feast and famine. But most available data suggest that they were surprisingly healthy and had a fairly stable diet and lifestyle. Not so the primitive farmers. In years when the crops failed, in settlements where the population density was high and where disease weakened the ability to cope even further, life would have been very hard indeed. The settled population could not migrate to follow the food supply as could hunter-gatherers. They were trapped.

With larger population groups and the further apportioning of skills, social stratification appeared. Power structures developed which had hierarchies, and these brought even more skill specialization: kings, soldiers, scribes, traders, tool-makers, priests, weavers, etc. Specialization necessitated living in larger networked communities, because there is no sense in someone specializing in making pots, for example, unless there is a sufficient demand for them from a sizeable community. And the pot-maker will depend on other specialists to ply his trade, for example he may need provision of pigments to decorate his pots, or even someone to take them to markets for sale. We can see how a simple interdependence of specialized labour can build up. We can also see how a system of barter—'I'll make you a pot if you milk my cows as well as your own'—will become impracticable as the complexity of society increases. Around 4,000 years ago, at the time of origin of cities in the ancient world, metal tokens of receipt for grain deposited in granaries started to be used, the basis of modern coinage,[29] it is possible that writing largely evolved as a way of bookkeeping. Many of these cities were located at strategically important sites, especially those connected with trade, because this becomes an important aspect of the economy of a large population when specialization and social structure develop. But as networks grew, so did the capacity for cultural as well as material exchange. This knowledge exchange accelerated the collective learning of a community and this in turn drove more technological development.

City life

Cities grew remarkably rapidly. Based on census records and on records about the needs of the population for grain and water, it is thought that the population of Rome had reached nearly half a million in the first century BCE, and that it had risen to nearly 1.5 million by the fourth century CE.

Some of those citizens lived in the Roman houses that we associate with the period, decorated like villas with small gardens and pools, and many slaves.[30] But there may have only been a thousand or so such houses. The remainder of the population lived in the estimated 50,000 'insulae' (or Roman apartment blocks) in the city and its suburbs, with many families occupying each building in cramped and squalid conditions—not so very different from slum areas of many cities today. The lower levels were often shops, taverns, or storerooms. These buildings often collapsed and the risk of fire and disease was high.

Archaeologists delight in the system of sewers of ancient Rome, which were first constructed in the sixth century BCE and extended considerably as the city grew. But for the urban poor at least, the water and sewerage systems did not connect to the insulae and in any case there was little sanitation above ground-floor level. Whilst the rich in their homes might have a lavatory, perhaps even flushed by a channel from an aqueduct, this was a rarity. Other citizens had to be content with shared toilet and bathing facilities, sometimes on street corners, which must also have been highly effective ways of disease transmission.[31] There would be raw sewage and even corpses on the streets—one record[32] has it that, while the Emperor Vespasian was at breakfast one day, a dog from the street brought in a human hand and deposited it under the table; there may have been much discussion about whether or not this was an auspicious sign.

So life in cities changed human existence. Crowded living in close association with animals promoted infectious disease and hygiene was poor. Rooms about 10 square metres could house an entire family on some fourth-floor levels in Rome. The food supplies became more unstable when there was crop failure—indeed there are many recorded famines in the ancient Greek and Roman world. Dietary balance changed, with little lean meat, fish, or fruit and vegetables for much of the population. Carbohydrate intake started to rise. There is clear evidence that health became increasingly compromised. Adult stature started to fall, reflecting the effects of infection and undernutrition in childhood. After all that *Homo sapiens* had done to change his environment by the use of ingenuity and the development of technology, now the environment was proving to be unhealthy. It was becoming outside the range for which our biology had equipped us. We had created environments in which we could still reproduce, but the price of living with the mismatch was becoming obvious.

With the development of settlement came organized bureaucracy (and taxes) and institutionalized religions.[33] The institution of formal marriage, even though not necessarily monogamous, probably appeared at the time of settlement, although the social structures that preceded it remain highly debated.

Urban populations continued to grow throughout many parts of the world. For example from the year 1000 until the Black Death in the middle of the fourteenth century, the population of Europe increased from about 30 to about 70 million—it fell dramatically after the Black Death by about 25 per cent. Surprisingly simple technological innovations made this rapid population increase possible. For example, from the eleventh century there was a widespread shift from a two-field to a three-field crop planting and rotation system. This simply meant that a population growing crops on half of the available land, and leaving the other half fallow for a year to recover, now moved to leaving only one third of the land fallow at any one time: where before only half the land yielded crops in any year, now two-thirds of it would. Clearly this was an enormous advantage, but it also necessitated an increase in the labour force and in specialized trades such as smiths and carpenters. Mining also became important because there was an urgent need for farm implements made from iron.[34]

The living and employment conditions of much of the population at this time were very poor. In the eleventh century it is estimated that 10–30 per cent of the workforce were slaves. These people naturally aspired to freedom, and with the advent of a more feudal system many were able to move to populate new villages: now they were serfs rather than slaves, working as indentured labourers or paying tithes to their master, but it is doubtful whether their conditions were much better. No wonder that many of these serfs sought the opportunities which city life seemed to offer: true wages for the first time and a degree of legal status. The ancient proverb 'Stadtluft macht uns frei' ('Town air makes us free') encapsulates this. We know a little of the quality of the air they breathed and the conditions in which they felt free—they seem horrifying slums by today's standards. Nevertheless there was a continual expansion and migration of population to towns and cities during the medieval period, and history shows the establishment of substantial markets for the sale of specialized wares, and the building of civic buildings such as churches, guild halls, jails, and even hospitals. The traffic in these urban areas, especially on market day, must have seemed no less congested than today's grid-locked cities.

Hard times

Charles Dickens wrote his novel *Hard Times* in 1854. He set it in the fictitious Victorian industrial town of Coketown. Smoke belches from the chimneys of the workers' cottages, huddled together along badly drained cobbled streets which gave no relief by having even a few trees. In fact to call the squalid housing of the day 'cottages' seems to show an economy with the truth of which an estate agent today might be proud. Most families did not occupy an entire house, rather a room or two in a building with a communal yard. These workers had moved to the city from the country, in part because of poor agricultural conditions and the penury of indentured labour to landlord farmers, in part perhaps because they genuinely thought that city life would be better.

It was not uncommon to have ten children in the family. So many mouths to feed meant that for working-class families children had to work as soon as they were able, usually around the age of 8 years. Family sizes only started to fall when schooling was made compulsory, creating less economic incentive to having many children. The crowded conditions of life, the poor sanitation, and the polluted air resulted in constant infections. While the energy demands were high, the diet was poor, and the average caloric intake was only about 2,200 calories, of which most was poor-quality bread and animal fat. Lean meat, vegetables, and fruit were rare components of the daily diet—just as they had been in ancient Rome. Child and infant mortality was high and this was reflected in an average life expectancy of about 30 to 35 years.[35]

But what about way of life for those peoples across the globe who had continued to be hunter-gatherers, pastoralists, or subsistence farmers in small proto-states until they were colonized? When this happened their lives were suddenly disrupted—in many cases the first contact proved disastrous, with the ravages of European infections such as measles and smallpox decimating populations. Even larger proto-empires such as the Aztecs were subject to rapid destruction by the military superiority afforded by the gun and the horse, if not by disease. The environmental changes for these populations were rapid, and tragic. From living in relatively stable social and physical environments, from the hunter-gatherer Bushmen of southern Africa to the large highly structured communities of the Andes, malnutrition, infection, dispossession, and sometimes slavery and displacement to very different physical environments became the norm. Peoples whose spiritual beliefs were still very much linked in time and space to their land, such as the Aborigine and Maori, found themselves displaced from their points of identity and identification. In a book of this nature it is not possible or appropriate to describe the enormous range of rapid environmental and emotional shifts such peoples faced. But the net effect was that the environmental changes which Europeans had experienced and adjusted to over 100 to 200 generations were faced by colonized peoples in five generations or less. The opportunity for adjustment, both biologically and through societal change, was thus much less. Many such societies are still paying the price.

Recent times

In the middle of the eighteenth century life was hard for most people in Europe. Samuel Johnson wrote that 'many complaints are made of the misery of life: and indeed it must be confessed that we are subject to calamities by which the good and the bad, the diligent and the slothful, the vigilant and heedless are equally affected'.[36] In 1759 the archetypal cynic Voltaire satirized his fictional character Dr Pangloss for declaiming that 'all is for the best in the best of all possible worlds'.[37]

By then Europe was reeling from the devastation of the Lisbon earthquake, estimated at force 8.7, and the three great tsunami waves over 13 metres high that followed, which killed between 30,000 and 70,000 people on All Saints Day 1755 when many of them were attending church.[38] The disaster seemed to call into question the idea of a benevolent deity who had a special regard for humans and would look after us. The world we lived in was changing fast—the new ideas of the Enlightenment were beginning to affect not just religion, but science and social structure too. And it has changed faster and faster, at an accelerating pace, ever since then. More and more humans have moved away from leaving their lives to chance, or from propitiating a deity which they felt would improve that chance.

The Enlightenment gradually spread across Europe in the latter part of the eighteenth century. Political change had been given impetus by the American and French revolutions and gradually democratic parliamentary systems appeared across Europe. The concepts of individual rights and freedoms grew. Some religious diversity appeared with the rise of less institutionally and more personally focused forms of Christianity such as Methodism. Agnosticism and Rationalism became more accepted. The scientific method of logical analysis and thinking was increasingly developed. Literacy increased; newspapers and books assisted emancipation by disseminating knowledge. Awareness of the greater world increased, especially as a result of information brought back from the expanding colonies.

During the mid-nineteenth century the dreadful conditions in the industrialized cities started to be noticed. This was after all a period of enormous development of science and technology. Many other disciplines can trace their origins to this period too—from anthropology to sociology and psychology. The spirit of Empire associated with Britain at this period suggested that there was no challenge too great, no part of our world that could not be explored, conquered, and made comfortable for 'civilized' humans. So why not improve conditions at home, in the cities where the workforce laboured so hard for the production on which the Empire depended?

Some of the cotton or wool mill owners were mindful of the plight of their workers. They attempted to give them some basic education and health care, to restrict their access to the gin that seemed to give them the easiest escape, and to build chapels to redeem their souls. By the mid-nineteenth century some cities such as Glasgow started to give real emphasis to improving public health. Sir William Gairdner (1829–87) was made Glasgow's Medical Officer of Health in 1863. He remodelled the city's sanitary system and set limits on the number of people who could live in a house at one time. The Glasgow City Improvement Trust was established in 1866 and undertook to clear slums and to widen streets. These were all changes for the better. But they were not by any means universally adopted, nor were they without critics.

Gradually governments became aware of their responsibilities to their people—and the role and impact of the trade unions was critical in driving this change. This culminated, after two world wars separated by a depression—through all of which the working classes fared far less well than the upper classes—in the setting up of welfare systems in various western democracies in the mid-twentieth century. These attempted to address the inequalities in educational provision for children from different backgrounds. They greatly improved access to health care, making it free in most cases, and introduced the infrastructure for improving nutrition, for example by providing free milk to schoolchildren. The post-war growth and health of children in such societies as the UK, many European states, Canada and Australia increased substantially in a generation.

Measures of growth through childhood and of mature height are an index of social, and particularly nutritional, conditions. During the mid-nineteenth century the average height of working-class men at maturity was about 13 cm less than that of their middle- and upper-class betters—that differential has now been lost.[39] In the last century there have been rapid increases in the height of people in the industrialized and rich nations—only now tailing off in countries such as Sweden and the Netherlands. For example, the average height of young Dutch men increased from 168 cm in 1900 to 184 cm in 1997.[40]

Dramatic changes in growth were also observed in the children of migrants from poorer parts of Europe such as southern Italy to the USA in the early twentieth century. These children were noted by Boas[41] to grow to be significantly taller than their parents, reflecting the dramatically different nutritional and health care conditions in North America from those experienced by their parents as children in Europe. We still see big growth differences when people migrate from the developing to the developed world.[42]

Getting healthier

The shaman arrives at the patient's dwelling bearing his traditional instruments—gong and stick, rattle, finger bells, veil, divination blocks, and sword. He prepares an altar, lights a candle, covers his face with the veil, and puts on the finger bells. The relatives of the sick person confer with the shaman and kneel to implore him to act. He assents. During the ceremony he cannot communicate directly—he has entered another world to negotiate with the evil spirits who cause disease, and the person who beats the gong must pass on his messages. A diagnosis is made, and the shaman begins a ritualistic chant intended to relieve pain and suffering. In return, the spirits are offered the sacrifice of an animal.

Is this a scene from a late Neolithic village? In fact, we're in an ordinary suburban house in the USA at the beginning of the twenty-first century. When the Hmong people from South-East Asia arrived in the USA as refugees from the Vietnam war, they brought with them a culture which emphasizes the role of the shaman in healing—not of physical illness itself but rather of the psychological and emotional distress associated with being ill.[43] This is a fundamental human need.

The beginnings of medicine are linked very much to the evolution of religious belief. Healing gods became a focus for, and their priests the providers of, primitive medicines. Belief and healing became linked well before medicine had a scientific basis. However we can assume that Palaeolithic humans used a wide range of plants as remedies, many of which may have had some efficacy because they had been discovered on a trial and error basis. They had a rich natural pharmacopoeia of remedies from which to choose. For example willow bark contains salicylic acid, the active principle of aspirin, and has been used for many years in Europe to relieve pain and fever. There is a long history of the use of extracts of poppy seeds, containing opium, as an analgesic, and quinine from the bark of the South American cinchona tree was used by native Americans as a muscle relaxant long before its value as an anti-malarial was recognized. Many such remedies have been handed down through the generations and may form the basis of 'traditional' medicine in many societies. Recent field observation has shown that even chimpanzees use a variety of plants as medications.[44]

Modern medicine and public health have had an enormous impact on health and lifespan, at least in the developed world. We have now conquered many of the infectious diseases which claimed so many lives even in the early part of the twentieth century—tuberculosis and polio to name only two. Before the development of antibiotics such as streptomycin in 1943, over 25,000 people died each year in the UK from tuberculosis. When an effective polio vaccine was developed and widely used in 1955 the sight of wards in many hospitals filled with 'iron lungs' for patients with polio disappeared almost overnight. Humans in developed societies now live longer than ever before—the average life expectancy in the UK has increased by about ten years over the last generation. We now have far better early diagnostic tests for many cancers as well as rapidly improving treatments for them; in addition to better targeted chemo- and radiotherapy we may soon have vaccines against some forms of carcinoma. And even though the threat of pandemics from viral infections such as SARS or avian flu are still very much with us, we have better anti-viral drugs and the ability to produce and distribute effective vaccines for new strains of virus quickly.

There are many epidemiological examples which stress the importance of the everyday environment in setting the risk of disease: in Victorian England the risk of tuberculosis was higher in drapers than in grocers, perhaps because in the draper's shop the door was kept closed and gas lighting used in its recesses to illuminate the wares. On a wider scale, as we clean up the environments of our cities, we see fewer cases of the chronic respiratory infections that led to so much misery fifty years ago. We not only have cleaner air in the streets, and certainly not the London fog or the Los Angeles smog of the 1950s, but our homes, offices, theatres, and shopping malls are well ventilated.

As we improve our water supplies and sanitation, there are fewer acute gut infections; many people in the USA or UK today have no relatives who died of such things. Yet cholera, typhoid, and dysentery were once endemic in New York, London, and many other capital cities. We now think of these diseases as 'tropical'. The same applies to the diseases spread by insect parasites, mosquitoes, ticks, fleas, and lice. Since the discovery by Sir Ronald Ross in 1897 that malaria was spread by mosquitoes, we have made enormous progress, although malaria still affects more than 300 million people worldwide and still kills more than a million people every year.[45] The quinine added to the gin and tonic tipple so popular with the empire-builders of the nineteenth century provided a partial protection against the symptoms caused by the malarial parasite. But now that Bill Gates has pledged millions of dollars to the development of an anti-malarial vaccine, a new chapter in the story of our fight against this disease is beginning.

As a result of this progress in medicine, life expectancies in developed and many developing countries have increased dramatically. At the same time control over reproduction and, for many, expectations about family size have changed. Together this has resulted in dramatic changes in the age structure of many societies. While even 100 years ago the age pyramid was indeed very much a pyramid, with very few people living to an advanced age (and it would have been a much shallower pyramid 10,000 years ago), now the pyramid has changed its shape as the proportion of people living beyond the sixth decade has risen dramatically. The societal structure is changing from many young and few old, to many old and fewer young people. Issues about support structures and the question of who needs to support whom become pressing.

The reproduction revolution

This book is largely about biological processes of evolution, development, and adaptation. Given the central role of reproduction in evolutionary processes we need to consider how reproductive behaviour has changed. In doing so we must limit our comments to western societies, for this aspect of human behaviour is very culture-specific. In western societies there has been a recent and dramatic change in reproductive behaviour, one that we have created for ourselves through technological development. It is often said that the part played by women in the war efforts of the First and Second World Wars played a major part in their emancipation. The post-First World War years led to universal suffrage, to the improvement of social status and rights: these have been recurrent themes throughout the twentieth century, and arguably are not fully achieved even now, even in the West. The post-Second World War years were followed by a second wave of feminism that revolved around rights to employment and to control over reproduction. This second aspect has been particularly important.

Various forms of contraception have existed for centuries, some of them of rather dubious efficacy let alone safety. And there have always been abortions performed, again not without risk to the woman. Infanticide was probably a common part of family management in the hunter-gatherer and remained so in such societies into recent times. In such societies, it was not practical to have children too closely spaced—both mother and her other children would be compromised.

But the advent of the contraceptive pill[46] changed everything. It was very effective, safe (or so it seemed), and, most important of all, it was in the control of women. Now at last was it feasible for a woman to have control over the timing of her pregnancy, in terms not only of the choice of a partner but in relation to her age, career, and other considerations. Not surprisingly, and partially as a result, the average family size across the globe has fallen from 4.9 to 2.8 children per woman and in developed societies it has fallen to well below 2.0. In parallel the average age at which women have their first pregnancy has increased. In the UK the mean age of women at birth of a first child has risen from 23.7 years in 1971 to 27.1 years in 2004[47] and over 25 per cent of women have not had a child by the age of 35. It has been possible for societies such as China to restrict family size by legislation. The processes of conception and childbirth are now part of the highly regulated world which we have developed for ourselves.

The science of reproductive technology has not stopped at the development of contraception. It has provided methods to help conception in couples who were previously infertile, through *in vitro* fertilization and more refined techniques such as intra-cytoplasmic sperm injection (ICSI), in which the sperm is actually injected directly into the egg under the microscope, and genetic diagnoses on the test-tube embryo before it is implanted. This is the field—the industry some would say—of assisted reproductive technology. This aspect of human cultural evolution now divorces reproduction from Darwinian fitness. We have provided medical methods to extend and to preserve life beyond what would have seemed possible 100 years ago. We have also made it possible for couples, whether fertile or not, to conceive at almost any post-pubertal age by use of these technologies. In 2005 a retired Romanian professor gave birth to a child at the age of 66.[48]

Transitions

This necessarily brief account of some key points in the history of humans has focused on highlighting several major transitions. The first was the evolutionary transition to technologically competent, reflective, cultured humans who communicated with language, which started in the middle Palaeolithic some 50,000 years ago. The second was the transition from hunter-gatherer to agriculturalist and settlement-dweller which started some 10,000 years ago. This continued progressively to the development of urbanization, complex power hierarchies, and the growth of cities, states, and empires. It brought with it massive changes in the status of individuals, their health, and the application of technology to all aspects of human life. The third transition was the technological revolution in both agriculture and industry starting about 250 years ago. The fourth, which in some respects ran in parallel, is a consequence of the Enlightenment and the greater recognition of the importance of protecting the human condition. In some ways this extends to the emancipation and human rights movements of the twentieth century—battles which are not yet completely over. Finally we cannot ignore the newest transition to a knowledge-driven society by a highly networked world with its rapid information transfer systems.

Each of these transitions in no small way led to human-driven changes in the environment, to which humans themselves then had to adapt. The effects of these major changes in our environment, from the Palaeolithic to the present, have put increasing pressure on our capacity to cope. Our basic structure and most of our physiology was determined by selection pressures which existed long before the modern era, and this must determine whether we are matched to how we now live.

It is difficult to separate components of the environmental changes that our species has faced, because nutritional, social, and societal changes are all intertwined. But nevertheless in the next sections we will do just that, to highlight specific aspects of these self-created environmental challenges.

The nutritional and work transitions

Our only source of energy is food. And we use energy in lots of ways. More than 70 per cent of our total energy consumption is used just to keep the body functioning. For example, the adult brain contributes about 20 per cent to our energy consumption (yet is less than 5 per cent of our body weight), and in the newborn this is over 60 per cent.[49] We also consume much energy in repairing body processes and, as we described earlier, one major theory of ageing is that it occurs because the body limits its energy investment in repair in order to invest it instead in growth and reproduction. The other major way we use energy is by physical work and exercise. This is usually about 20–30 per cent of our total daily energy use, although it can increase enormously in people who engage in vigorous exercise. Growth, reproduction, and lactation are all additional and special forms of energy consumption where the nutrients absorbed are sequestered in new tissue formation or are transferred to the fetus and infant for its growth.

So on balance we generate energy by eating, and use it through growth, operating our bodies, reproducing, and exercising. Any excess energy is stored primarily as fat. The potential for different foods to generate energy differs. Fats are high energy supplies, carbohydrates and proteins poorer energy supplies. But the mere act of digesting, absorbing, and metabolizing different foods itself consumes energy. The field of nutritional science is in large part about understanding the ways in which the net effect of a particular food in a particular situation affects this energy balance.

There has been an enormous shift in this energy balance as we have changed our world. The Palaeolithic diet had a high-protein, low-carbohydrate content and the nature of the fats in it was different. But diets started to change with the development of agriculture and workloads started to rise too. With settlement many people became dependent on others and had to purchase or barter for food rather than directly controlling their own provision. Agriculture brought a reduction in food security with periodic famine induced by drought.

The second agricultural transition brought about by mechanized farming coupled to industrial food production, packaging, and distribution also had dramatic effects on the food–energy balance. In the twentieth century foodstuffs started to be enriched not just for nutrient value but also to make them taste better. Highly refined sugars were added to foods. Their net energy value was much higher than the traditional carbohydrates, which are digested more slowly and require energy consumption to utilize them. Consumption of higher-fat diets followed greater use of farmed animals as a food supply—farmed animals are usually fatter than their wild counterparts, because they too have to consume less energy to get their food and because there are economic incentives for the farmer to get his animals to gain weight fast, often in the form of fat. Fatty meat also tastes better—hence the high value placed on beef marbled with fat in Japan. The development of fast food outlets, the impact of media in supporting food fads rather than balanced nutrition, has become a runaway train. There can be no doubt that the quality of our diet now is very different from that which we consumed throughout the vast majority of our evolution.

While this change has been progressively accelerating in the developed world, the rate of change has been even more dramatic in the developing world where the 'nutritional transition' has been almost instantaneous. People in many such societies only two or three generations ago still ate a pre-industrial subsistence agricultural diet, or even a hunter-gatherer diet; now such peoples are exposed to increasing amounts of foods produced in the western style. This rapid transition is playing a major role in the epidemic of diabetes in countries such as India. Even in sub-Saharan Africa where the general level of nutrition is often lower there is evidence that obesity is greater in the urban rather than rural poor and that the incidence of 'lifestyle' diseases such as diabetes is increasing.

But as much, or even greater, change is happening on the other side of our energy balance equation. For if we expend less energy than we take in, we store the excess as fat. In traditional societies most people played a role in food gathering and this consumed energy—this was true for hunters as well as pastoralists and manual agriculturalists. With the development of towns and cities, the dominant individuals in the social hierarchy avoided hard physical work, but they were few in number and energy expenditure probably remained high for the majority of people. Their situation may not have been so very much different from that of the hunter-gatherer or it may have even been worse.

Indeed, contrary to popular understanding, the hunter-gatherer, while probably consuming more energy during the day, spent considerable time neither hunting nor gathering and this time was presumably used in social interaction.[50] But with the industrial revolution all sorts of 'energy-saving' technologies appeared—from the mechanical plough to the railway engine. In a very short space of about 200 years energy expenditure fell dramatically for many people. The task of obtaining food for ourselves and our families changed from being very energy-consuming to minimal. Now we can use the internet to order food to be delivered from a supermarket.

The development and widespread availability of cars meant that transport became much less personally energy consuming—although in terms of energy use on the planet there has been an enormous increase. It is clear that the appearance of childhood obesity in the past thirty years correlates with two factors: the increased use of cars rather than legs for transport, and the increased use of television and video screens for leisure rather than more traditional pastimes which often involved physical activity. Even in countries such as India childhood obesity is appearing in those children who no longer walk or cycle to school.

Social transitions

We can really only speculate about the social environment of Palaeolithic humans. Prior to the development of settlement, humans lived in social groups of less than 150 and perhaps as small as 20 to 50 people.[51] These would have been extended family groups, and this has been an important component of our species' success. For example there is evidence that child survival is higher where the maternal grandmother lives with the mother and child—she can help with childcare, cooking, and feeding and teach the mother essential mothering skills. Only occasionally would bands coalesce with other bands; perhaps for trading and for dispersing adolescents to avoid inbreeding.

With settlement, much larger numbers of people came into direct contact with each other. Those who lived in cities came into contact with many hundreds of people. From living in a small clan where individual roles and relationships were clearly evident and the power hierarchy simple, humans came to occupy complex networked social structures where roles were subdivided and separated and intricate power and control hierarchies emerged. More recently with the emergence of a nuclear family the risk of isolation from a social support network has increased and more people are lost in a bewildering complex of rules and decision-making processes.

The pace of technological change was minimal in the Palaeolithic and, at first, slow in the Neolithic. But as knowledge networks expanded and settlement became more organized, the pace of this change started to accelerate. In classical Greece, Egypt, Persia, and China and later in Meso-America, mathematical, philosophical, and engineering progress was substantial. The European post-Renaissance scientific revolution gave real impetus to science and technology (Galileo, Copernicus, Newton, Leibniz, all spring to mind) and scientific thought picked up further speed in the nineteenth century in response to the technological needs of the industrial revolution. Through the twentieth century we have seen an exponential increase in the complexity and application of technology, much of it affecting our lives directly. In one generation we have seen the appearance of the jet aeroplane, the rocket, the computer, the internet, television, credit cards, fax machines, microwave ovens, nuclear weapons; each one of these inventions provides something new to which we have to adjust.

There are serious questions about how much capacity we have to adjust to wave after wave of new technology. Superficially we seem to cope, but consider the change in response time in an interaction between the two authors of this book. Fifty years ago an exchange of letters between us would have taken twelve weeks. Then came airmail and it was reduced to two weeks, then the fax reduced transmission of a long document to minutes. With email we can send entire drafts of this book to each other almost instantaneously. Those involved in the study of business organization now recognize that this creates a demand for instantaneous responses which induces considerable stress in employees.

Mismatched in our world

Humans can be distinguished from all other species in the way in which they continually, and often intentionally, modify their environments, both physically and socially. For much of our history we employed our unique capacities of thinking, communication, planning, and use of technology in rather limited ways to expand the range of environments we could inhabit. When we stopped living as nomads we started using technology to intensify our population density in specific and static environments. The pace of that change has accelerated as technological developments have progressed.

Throughout our history as a species we have striven, by the use of our intelligence, our ingenuity, and by sheer hard work, to improve the conditions under which we live. It is better to be warm than cold; better to be replete than hungry; better to live for another year than to die today of disease; better to stroll to the refrigerator to prepare a sandwich than to walk for miles to collect a few meagre nuts and berries. Life really is so much better than it was, at least in developed societies, and only the most cynical person would deny it. We have improved the lot of the human condition for ourselves, our families and friends, and even for the particular society in which we live (our tribe if you like) beyond all recognition since we migrated out of Africa.

And so we have come to design sophisticated ways of growing and distributing food, and of manufacturing the foods we seem to like best. We have applied technology to reduce the amount of physical work most of us expend every day. We have developed a complex economic system that allows us to adapt to new environments by purchasing what is necessary in order to deal with them. We have developed measures to prevent disease, and to treat it if it occurs. We have promoted social structures to sustain the lives of more frail members of a population. We have explored ways to allow everyone to reproduce if they wish. Most of us now live for longer in a cleaner, safer world, and we have the leisure to enjoy it. But the impact of humans on the environment is increasingly apparent—as Polynesian islands disappear under a rising sea,[52] as polar ice caps melt and glaciers retreat, we can truly reflect on how we have changed our environment.

Every 'improvement' we made by the use of technology has only further changed the environment to which we had to adapt. We changed our nutritional environment, our disease environment, our social environment, our societal environment; we lived longer. As the degree and pace of change increase, so the question has to be asked—what limits in our biology might be exceeded by the environments we are creating and what are the consequences? This question comes into sharp focus because two aspects of our biology have *not* changed. They are the two histories each of us carries with us, from our evolutionary past and our individual developmental past. We discussed them in Chapters 2 and 3 in the context of the ways in which humans try to be matched biologically to their environment. If we had not tried to be matched, we would not have survived to this day as a species. But now we have changed that environment ourselves—changed many crucial aspects of it, and changed them very fast. The health and social problems which the Kanuri people faced when irrigation was introduced into their more traditional farming practices might be faced by many of us around the world, in different but no less important ways. Could it be that in trying to make things better, we have become increasingly mismatched to our environment?

5
Constrained by our Pasts

Perspectives on time

History and biology do not just work on a single time scale. An evolutionary biologist thinks in terms of tens of thousands if not millions of years, an archaeologist in thousands of years, a geneticist might think in terms of a few generations, a doctor or biographer in terms of a lifetime.[1] The mismatch paradigm tells us that we cannot shake off our biological pasts of evolution and development. It continually reminds us that the range of environments we evolved and developed to inhabit can be very different from those we actually live in now.

The shaping of our genetic repertoire occurred as our species evolved and dispersed around the globe. Our ancestors faced the challenges of diverse physical environments and also their own impact on these environments: these in turn affected their sources of food, their exposure to disease, and the social grouping and structure within which they lived. The more recent part of our individual journey started when our mother was an embryo and the egg destined to make each of us formed in one of her developing ovaries while she was still inside our maternal grandmother's womb. We have seen how the environment of that egg as it was fertilized and grew to be our mother, who in turn incubated each of us and created an environment in which we grew, has influenced us. After we were born other factors influenced our biology in many ways. These environmental messages from our many pasts have set *constraints* on our current biology.

Although we might wish to, we all eventually come to realize that we cannot do everything we want to in our lives. We are bound by a wide variety of constraints. Virtually everyone faces a significant financial constraint which limits their ability to indulge in their fantasies, e.g. of buying an island or having a private jet. And there are social constraints on how we behave. Our sexual mores are determined by the society we live in. Our families, friends, jobs, and social structure create emotional and other constraints. And we have obvious but poorly understood biological constraints which prevent us living forever. Just as we inherit both financial and social opportunities and constraints, so it is that our various pasts—evolutionary and genetic, developmental and epigenetic, environmental, cultural—from long ago and from more recent times—create opportunities and constraints on how we can live healthily. We have implied in the last three chapters that these constraints exist. Let's now be more explicit about them and bring our ideas together before discussing, in the second part of the book, how the resulting mismatches affect our lives. While we might conveniently think of constraints separately, they are really intertwined like the strands of a rope.

The first strand: evolutionary constraint

The fundamental processes of evolution operate to match the organism to the environmental range it inhabits. Evolutionary processes work by selection of characteristics which confer a reproductive and survival advantage. Key to the processes of evolution is the generation of variation in the gene pool which in turn is driven by mutational processes. But there are constraints on what variation can be selected.[2] Some genetic changes will interfere with basic body processes and are incompatible with life. The effect of many other mutations will be covert and have no observable effect, although an effect may be uncovered in a different environment. The effect of any one change may be limited by the design features of the organism. And there may be genetic variation in genes that regulate the expression of other genes—this is seen as one way in which a coordinated change in a range of body structures, functions, and behaviour can be induced by a change in a single gene.[3]

Increasingly we recognize that much selection is not necessarily for a characteristic *per se* but for the capacity of the individual to change that characteristic in response to an environmental change. For example, although all fish originated from a common ancestor some 400 million years ago, some are adapted for salt water and some for fresh water. All fish must maintain their internal composition constant or else they will die, so fish living in the sea have different set points for their salt and water regulation from those living in fresh water. But some fish can live in a broader range of salinities than others (some salmon and eels migrate from fresh to salt water and back) and these must have evolved by selection to have a capacity to regulate salt and water balance over a wider range than fish that can only live in salt or fresh water. Through these processes all species become matched to an environmental range; for some (specialist) species the range is very specific and narrow, whereas for other (generalist) species it is broader.

Humans have evolved particular attributes including a large brain, the capacity to plan, communicate, and use technology—this makes us the ultimate generalist species because we have the ability to change our environments. But even so, there are limits to our adaptive range. These limits were established during our evolution, but the environments we increasingly create for ourselves are different. To have a wardrobe full of tropical clothes might be ideal if we live in Hawaii; it is rather problematic if we find ourselves living in Anchorage. So it is with selection—it has given us a repertoire of internal clothes to allow us to live and breed within a range of environments. But the wardrobe was chosen at a different time and in a different environment.

So humans are characterized by a greater ability to control their environment than any other species. This capacity was appreciably and progressively enhanced with the technological explosion that came first with the Bronze Age and has progressed exponentially to our nuclear and electronic age. As the environment became more controlled, the need to respond by evolutionary selection consequently declined. No longer did the human species have to change its gene pool to adapt to the environment, it started to change its environment to match its gene pool—we no longer had to rely on our biological internal clothes to make our lives matched to our environment. We started to go shopping for new external clothes. This does not mean that evolutionary processes no longer exist, merely that shopping became preferable (it still is—ask our daughters).

The potential for Darwinian evolution in humans remains—certainly molecular studies of the human genome show that changes have occurred in the DNA sequence of some genes within the last 10,000 years and some of the genes reported to be altered are associated with nutrition and metabolism.[4] But if this reflects natural selection, much of this change was likely to have been associated with the transition from hunter-gatherer to lifestyles associated with farming and settlement several thousand years ago rather than being due to recent events. Indeed many of the pressures that would previously have driven natural selection in humans have been relieved in our more recent past.

Few human populations are now truly isolated. We generally respond to changes in the environment by manipulating our environment to allow us to cope—from astronaut's suits to housing to clothing, to medical technology and medicines, even to food aid—these are all means used to change our environment. We increasingly change reproductive fitness by allowing most who wish to reproduce to do so, at least in western countries. So we provide assisted reproduction technologies to those who are having problems conceiving. Human evolution has continued but in many ways it may have been dampened by our interventions. Yet selection pressures are likely to be almost as important now in our responses to infection as they have always been—they are only modified to the extent that medical treatment can keep alive those who would otherwise have died. The risk of a pandemic remains high, as the recent concerns over avian flu demonstrate. For all we know, there may well be genetic determinants of who can survive the next pandemic, as there appear to be to an extent for AIDS. If the virulence is sufficiently high to kill young people or those of reproductive age in large numbers, such an epidemic could alter the gene pool—this would be evolution at work. And there may be new challenges that could arise, for example from a nuclear winter or global warming in which only a few individuals with an appropriate phenotype might survive.

There is indeed evidence to show that our capacity to fight infection has moulded how the human gene pool has evolved. Sickle cell anaemia is a disease caused by a genetic variation in the structure of haemoglobin, the protein in our red blood cells which carries oxygen. If a person has two abnormal copies of the haemoglobin gene, they develop major problems with the quality of their blood cells and they suffer all the effects of severe anaemia and its complications. But oddly, if a person has only one copy of the abnormal gene they are more resistant to malaria. It is thought that this is the reason why the mutations leading to the sickle cell trait are so common in the gene pool in sub-Saharan populations where malaria is endemic. About 8 per cent of African-Americans carry a single copy of the abnormal gene and are carriers of sickle cell disease.

So while human evolution is not 'dead' as is sometimes suggested, the reality is that how we manage our environment rather than changes in our gene pool will almost certainly determine our foreseeable future.[5] This is a fundamental constraint on what lies ahead of us.

The second strand: design constraints

Part of our evolutionary inheritance has been to acquire the tools of developmental plasticity which allow us to adjust our developmental trajectories, and thus our life-course strategy from early life onwards. This capacity has obviously been fundamental to successful life on this planet, and indeed many organisms—plant and animal, from single cell to the most complex—have a plasticity toolkit. A major motivation behind writing this book is to describe how important this particular toolkit is, and how we can use knowledge about it to our future advantage.

Developmental plasticity extends the range of environments we can adapt to. It has two fundamental components—a sensor and a responder. This is just like a police radar trap for speeding drivers. This sensor is the policeman checking speed with radar as we go past. He then sends a message to the responder, saying, 'Book that one'. The responder might be the local authority that sends a speeding ticket through the post or it might even be another policeman on a motorcycle further down the road who pursues us. Our environmental sensors must work in multiple modalities as developmental plasticity is about all the factors in the environment that might affect life and reproductive success, and about which the organism must try to do something. These environmental factors can include competition for food, danger from predator species, conflicts within the species for food or sex, and so forth. Our biological responders must also be complex and depend on the system being evaluated by the sensor, just as the responses to a speed detector and to a smoke detector are very different. The response system is made more complex in biology in that the responses invoked can take effect either immediately or later in life.

But developmental plasticity has one particular and critical component—for most organs it is essentially irreversible and once a path is chosen we cannot go back. This constitutes a very fundamental constraint. For example, the number of kidney units (nephrons) is determined in fetal life and this number can be influenced by nutritional and possibly hormonal factors from the mother. If a kidney is damaged later in life, each of these units may try to increase its function but no more units can be formed. The brain is an organ with some components irreversibly influenced by developmental plasticity and others which remain relatively plastic throughout life. The location and number of brain cells is largely determined in fetal life but many more neurones than are ultimately used are formed. Many die, and the survival of the others depends on their connectivity and activity within neuronal networks. For example, in the infant cat, covering one eye for a critical period in the first three to six weeks of life, so there is no stimulation of the visual pathway as it is developing, changes the pattern of connectivity that forms within the brain and leaves the cat with permanent monocular vision—a discovery that contributed to the award of the Nobel Prize in Physiology or Medicine to David Hubel and Torsten Weisel in 1981.[6] In contrast, learning is a form of plasticity which, while it has much greater capacity in early life, can nonetheless continue to a degree throughout life—youngsters can learn to play the piano or speak a new language much faster than middle-aged people but even older people can learn these skills to some extent.

Developmental plasticity can be mediated through changes in either structure or function or both. Many of these changes in turn depend on environmentally induced epigenetic modification of the DNA so that gene expression is altered. But there are even further constraints. Whatever developmental response occurs in relation to the environment, the change in the organism must be viable. If the tadpole can develop either a short or a long tail, whatever tail length it develops must allow it to swim and to feed and be compatible with future metamorphosis into a frog. The need for all options chosen to be viable creates an essential design constraint.

Another constraint is determined by the practical point that the nature of the response depends on when in development it is induced. If you are building a house you cannot go back and change the shape of the foundations when you are putting the roof on. And you cannot put the roof on until the walls are complete—there is an order in building and that basic order cannot be changed very much. The sequence of events also matters in development—once the limbs are formed they essentially cannot be remoulded, the fingers cannot develop until the hand is formed, and so on. Just as there would be an enormous cost to trying to reshape the foundations of a house once the walls and ceiling are in place, so there would be a very great biological (and hence fitness) cost to retaining the capacity to be infinitely plastic in every respect throughout life. The more differentiated and specialized the cells have become, the more complex the organs, the more complete the physiological control settings, the harder it becomes. Thus for each biological system and for each component within it there is a critical window of time within which plasticity occurs and beyond which it is no longer possible to change, because the costs and practical design constraints would just be too high.[7]

But all this assumes that the sensors are working well. If the smoke detector is faulty we may keep evacuating the house after false alarms; or worse we might be suffocated if it doesn't sound when it should. The larva in a cocoon, the embryo in an egg, and particularly the fetus in the womb have a very limited ability to sense the outside world directly. Nearly all the information the fetus gets about the world out there comes to it by way of signals from the mother. She acts as the sensor and the signals to which the fetus responds are indirect.[8] There are many opportunities for the information to be corrupted before it reaches the fetus. Thus the fetal response may become inappropriate to the environmental signal, and this is not without consequences.

We have suggested that the developmental response to an environmental signal can be short- or long-term. Short-term responses may be necessary to respond to immediate threats to continued development. Longer-term responses allow the individual to optimize its development and life-course strategy to give it the best chance in the adult game of 'food and sex'. It is important to understand that these long-term responses are balanced and integrated. We have evolved with a toolkit which allows us early in development to tune our adaptive responses so that they match our predicted future environment. In evolutionary terms the ultimate goal is to increase our reproductive fitness. So if we predict hard times, we have evolved to use a strategy involving less investment in growth, development, and tissue repair, and a preference to reduce energy expenditure by lowering activity and favouring energy storage as fat. Conversely, if we predict that times will be easier we will favour a strategy in which we maximize our growth and plan for longevity. Where we are between these extremes depends partly on how we read (or misread) our mother's environment as a fetus, and how we used this information to predict our future environment. This is the way a match is induced—or alternatively developmental mismatch arises.

An important issue in the use of short-term responses is whether they have costs that will have to be paid later. Often they do. Being born small may be a short-term response to a poor fetal environment with limited nutrient supply. But a newborn animal that is born small is more likely to die, to lose in the competition for food with members of its own species, or to be predated. In humans, smaller babies are more likely to die in the period around birth and the smaller we are at birth the greater the risk. But the consequences are not restricted to the neonatal period, because as we shall see there is a price to pay that extends throughout life.

For longer-term responses the key question is whether the choices made on the basis of developmental forecasting are correct. To the extent that predictive responses determine the developmental trajectory and later adaptive capacity, the fidelity of the prediction creates a major constraint. The key question is how good the fetus is at predicting the environment outside the womb. It can be misinformed because of detector problems but it can also be influenced by what it has inherited. If the mother is of short stature because of illness, or if she is very young or very old, then there will be erroneous legacies passed on to the next generation. And we increasingly recognize that environmental influences in one generation can induce epigenetic change that can be transmitted over several generations.

The fetus is totally dependent on its mother both for its nutrition and for sensing the state of the nutritional environment into which it will be born. But there is a very complex pathway from the environment that mother lives in to that the fetus senses. While the fetus ultimately detects those nutrient signals from the nutrients it takes up into its tissues, what gets there is determined in part by the diet, health, metabolism of the mother, and her physiology which controls the supply of nutrient-containing blood to her uterus. Then there is the state of the placenta itself—is it functioning optimally in transferring nutrients from the maternal to fetal circulation, and how many of these nutrients does it burn up to meet its own energy needs along the way? So, like a game of Chinese whispers there are many points at which the message can get garbled. The nutritional information the fetus gets may well not be an accurate reflection of the nutritional environment to which the mother is actually exposed. This is a design constraint—developmental plasticity in the mammal depends on this rather inefficient sensor system. And aspects of maternal behaviour such as smoking and unbalanced diet can interfere with the process, as can maternal ill health and placental malfunction. This is the reality of mammalian development—it is not a perfect process.

The third strand: the constraint of birth

Our success as a species depends primarily on our large brain. But the shape of the pelvis has had to be altered to allow our hominid ancestors to walk on two feet. The pelvis had to be changed in dimensions and in the position of the hip sockets, narrowing the width of the pelvic aperture. But we have a much bigger head than our primate cousins because of our larger brains. The only way out of the problem was to evolve to be born in a very immature state relative to other primates and to leave more of our brain development until after birth. The baby chimp is fully mobile from birth; we have to wait a year before we can even start to walk. While we are born prematurely with respect to brain growth, the fit of the human baby's head through the pelvis remains very tight, in contrast to the very easy passage of the small mature fetal chimp's head through its mother's relatively wider pelvic canal.[9] Being of the right size to allow us to be born is our key developmental bottleneck. If fetal growth was controlled only by genetics the problem would be even greater. Think of the scenario of a genetically large male mating with a small female. Without mechanisms to override the paternal genetic influence on fetal growth, the resulting fetal overgrowth in relation to maternal size would be an evolutionarily very dangerous situation—the fetus would certainly die and probably so would the mother.

The way human evolution appears to have resolved this potential conflict is to magnify the importance of mechanisms already existing in other mammals—namely to exploit the processes of maternal constraint.[10] These complex and poorly understood mechanisms make the control of fetal growth in the second half of pregnancy largely dependent on the supply of nutrients from mother to fetus. Maternal height and pelvic size are correlated, and in turn pelvic size and the size of the uterus and its blood supply are also correlated with each other and with maternal size. This means that maternal size will determine in no small part how much nutrition can get to the fetus. Thus fetal growth is closely linked to maternal size. This permits the human fetus to limit its growth not by its genetic potential, but in relation to its intrauterine environment. Thus the fetus can usually successfully pass through the pelvic canal and another generation is born. We believe this mechanism is critical in understanding how developmental mismatches involving nutrition arise, because maternal constraint effectively limits the messages to the fetus about the nutritional state of its future environment.[11]

This constraining influence of the mother extends beyond birth. Infant humans, like most mammals, are entirely dependent on their mother for nutrition, but in humans this dependence even extends beyond weaning. Weaned infants still do not have the skills to forage for adequate food or to care for themselves. Weaning is thought to have occurred between the ages of 2 and 4 years in Palaeolithic humans as it did until recently in Australian Aborigines prior to European influences changing their way of life.[12] This seems a very long time by today's standards, but in prehistory suckling aided survival of offspring because it extends the period of maternal support by expanding the birth interval.[13] But well before weaning supplementary foods may have been introduced. In modern humans we know that prolonged exclusive breast feeding beyond 6–12 months is associated with a decline in nutrient quality. We also know that the quality of breast milk is influenced by the mother's health. Thus under poor conditions nutritional influences constraining development extend well into infancy—we will see how important this might be.

Weaving the strands

We have described in the first part of this book how we evolved to try to match our biology to our environment to live within a comfort zone. However we have intrinsic constraints on how far we can adapt. Some of these constraints are genetic in origin, others are developmental. But humans also have a remarkable capacity to change their environment and we have been responsible for some enormous changes in our environments in the last 10,000 years. As we confront these changes in our environment, our capacity to respond is constrained by our various pasts. Our evolutionary past tries to match the range of phenotypes we can develop, using our repertoire of genes, to our environment—but that environment was largely determined more than 10,000 years ago. We have also been equipped through our evolution with the toolkit of developmental plasticity. This allows us to tune the degree of match with our environment further but there are limits imposed by our design. Together these processes define the range of environments we can live in healthily. But there are boundaries to any comfort zone. The Sherpa and the Kanuri crossed these boundaries. And we increasingly challenge these limits as our society and our environments change at a rapid pace. The greater the degree of mismatch, the greater the cost, and the more we need to understand it.

PART II
Mismatch

In the second half of the book we reflect on how humans live in their environments and how the combination of genetic and developmental constraints can affect our lives. Do our new environments, which we ourselves have created, match the range of adaptive processes which we can express? If not then the risk of greater mismatch must have increased and this will be manifest in altered disease risk. What are the consequences? If we were to search for evidence of the effects of mismatch between our inheritance and our environment where should we look?

In Chapter 6 we examine the maturational mismatch which has arisen as our society has become more complex and children get healthier. The coincidence of physical and psychosocial maturation at puberty which existed throughout most of our history has been lost, and we will discuss the profound consequences for modern society. They challenge the very way we think about adolescence. In Chapter 7 we look at the consequences of the dramatic changes in our diet, food intake, and energy expenditure. This has led to increased levels of metabolic diseases such as diabetes and obesity, and to linked conditions such as cardiovascular disease. In Chapter 8 we discuss middle and old age, because selection can only operate until the end of reproduction and humans now live longer lives. Developmental trajectories selected in early life may not be advantageous in later life. Mismatch can thus arise as the consequence of living much longer and this gives a different perspective on ageing, the menopause, and diseases of old age.

These scenarios are the starting points for the next phase of our journey of discovery. We must confront the consequences for our health as individuals and for our society. In the final chapter of the book we bring these ideas together to consider the mismatch paradigm in perspective, to think of ways to ameliorate the effects of mismatch and thus how to improve the human condition.

6
Coming of Age

Soon after girls of the Masai tribe in Kenya experience their first menstruation they are married. But boys from the same tribe, having spent the first twelve years of their life tending the cows and goats, are diverted into their 'warrior period' as soon as they enter puberty. With much ceremony and dances evoking their future roles as warriors, they are separated from the rest of the village. They will now live together, eat and sleep together, learn together, and fight together. Their role shifts from juvenile cowherd to tribal warrior and they now have to protect their tribal lands from intruders. After about ten years as a warrior and with the same great ceremony they experienced when they entered the warrior class, they then leave it to become 'elders' (at the grand old age of 25). Only now do they marry the much younger pubescent girls and settle down. As Robert Sapolsky puts it, 'perhaps they then spend their days complaining about the quality of today's warriors'.[1]

On some small islands of the Gulf of Carpenteria in the tropical north of Australia, the males of one indigenous tribe are similarly sequestered at puberty at which time they are circumcised. Then, instead of speaking the language they grew up with, they are taught a special sign language which they only use during their initiation year. Then they face a second genital mutilation—subincision (a crude operation to change the position of the urethral (urinary) opening from the tip to the underside of the penis)—and learn yet another, this time spoken, language. This uses quite distinct sounds and has a different structure from their childhood language and only the initiated men can use it.[2] In most Aborigine societies prior to European contact, as in many other hunter-gatherer societies, girls were married at about 14 as they completed puberty while men had to wait until 25 to 35 to receive their first (and perhaps only) wife. In part such a system developed to cope with the high neonatal mortality and the problem of an abnormal sex ratio generated by selective infanticide of females used in such societies to try to match the population density to the capacity of the ecosystem to support the clan.[3]

Puberty rites are common across cultures, particularly in traditional societies,[4] and serve to mark the transition of child to adult. They provide an important period of cultural learning and are very society-specific. There are echoes of such ceremonies of life-course transition in the communion of the Christian faith, in male circumcision in the Muslim faith, and in the barmitzvah of the Jewish faith, which confers the status of adult on the 13-year-old boy.

Such rituals in traditional societies usually involve separation of the adolescent from his or her parents, marking the transition from parental dependency to independence. Sexes are segregated and often secluded, then instructed in adult gender roles and in elements of tribal wisdom. For males the rituals may be very prolonged and involve pain and brutality, tests of strength and courage and role-modelling as a warrior. For the female they tend to be shorter and focused on the roles the woman must soon take because marriage for the female almost always happens soon after. The learning requirements for females, who will have already acquired many mothering skills while assisting their own mothers or aunts, are generally perceived to be less than for the male who is inducted into the tribal roles as warrior and hunter.

The timing of these events is informative. For girls these rites are always linked to their pubertal maturation and they generally enter marriage immediately after puberty. For boys, the timing is more variable, in part because the markers of puberty are far less distinct and in part because in many cultures young males are not allowed to have full reproductive rights until they are much older.

The period of biological maturation and the transition to assuming the responsibilities of adulthood are tightly linked, particularly in girls in traditional societies. But in western societies, rather than being a distinct period with both biological and psychosocial components, the transition from child to adult is a long and sometimes arbitrary process of accumulating rights. The timing varies somewhat between countries but follows the same general pattern. For example in New Zealand at 12 one gains the 'right' to be prosecuted for a crime, at 15 the right to drive a car, at 16 the right to buy tobacco or have sexual intercourse and get married, at 18 the right to join the army and to vote, at 20 the right to buy alcohol, and at 25 one is allowed to hire a car.

Put in this way, this long-drawn-out transformation, extending well over a decade, seems illogical and in fact it is a recent phenomenon. Just three or four generations ago puberty marked a relatively rapid transition from child to adult—be it as a wife or as a worker. For women, marriage occurred in the teenage years; but many teenage boys such as our grandfathers left home, even emigrated without their parents, and worked quite independently. Now it is becoming increasingly apparent in western countries that adolescence and some level of dependency extends for many into the third decade.

At the same time the western world has become very aware that the age of physical maturation is getting earlier and earlier. Almost half the girls living in countries such as Spain and Italy have started menstruating by the age of 12. It was not always this way—their grandmothers would have on average started menstruating at 14–16. Is this earlier physical maturation a normal or an abnormal (i.e. pathological) event—is this a phenomenon that can be understood from the evolutionary context? Is something frightening going on, for example as a result of our exposure to environmental toxins? And what are the consequences of this early maturation? This is the topic of this chapter. But because of the very important cultural differences in how adolescence progresses between societies, with very different patterns of adult acculturation, our discussion can only focus on modern western society and the turmoil its adolescents face.

Growing up

For some animals reaching reproductive competence and ensuring the passage of their genes to the next generation is not only the most important goal of their lives but also marks their impending death. The Australian redback spider will become food for his mate—sacrificing his body to her appetite after mating. Many insects spend most of their lives as larvae or caterpillars, and have only a very short existence as an adult after metamorphosis, when they quickly lay or fertilize their eggs and then die. The same is true even for some vertebrates. The salmon who struggle upstream to the place where they were born in order to lay their eggs or sperm pay the ultimate price for this incredible feat of navigation and physical strength: they are so exhausted by the effort that they die. This strategy of total investment in a single reproductive season—called semelparity—even occurs in some male marsupials. The marsupial mouse *Antechinus stuarti* is a small and mainly nocturnal carnivore which lives amongst the logs and leaves on the ground of south-eastern Australia. The male only survives for about a year while the female has a lifespan of about five years. At about 11 months of age the males enter a mating frenzy where they copulate as often as they can. Some die fighting other males but many perish from weight loss and infection brought on by the stress of their sexual excess. Within a couple of weeks all the males are dead. The benefits of such an extraordinary life history, and how it evolved, are unclear.[5]

But most mammals can reproduce more than once although there are some fundamental differences between the strategies employed. Some species have relatively short lives but breed 'like rabbits' during them. So rabbits, mice, weasels, rats, and voles have evolved a 'fast and furious' strategy based on having many progeny and gambling on a few surviving to breed. Their young mature fast and can reproduce at an early age, very soon after weaning. The female rat has a pregnancy lasting twenty-one days; her pups are weaned at twenty-one days and have completed puberty by fifty-five days of age so they can reproduce soon after. After weaning her offspring, the female rat makes no continued investment in them—it is more important for her to get pregnant again and to leave her juvenile pups to fend for themselves. She can be a grandmother within five months of being born herself.

Humans and the other large mammals such as elephants, horses, and whales have a very different strategy. They grow and develop slowly, have long lives, and invest an enormous amount of resource in the very few progeny they have, each of which they nurture to increase its chance of surviving to adulthood. Humans play this game of maturation in a particularly extreme way. We have a much larger brain to develop and, because of the challenge of the passage through the pelvis, we do more of its growth after birth than do other primates. Our brain size at birth is 25 per cent of adult brain size whereas the baby chimp's brain is nearly 50 per cent of adult size at birth. So we must be nurtured though a long infancy, childhood, and juvenile period before we are ready to encounter the challenges of the transition to adulthood.

It has been suggested[6] that our dependent childhood period is rather unusual in contrast to most other large mammals where the progeny move rapidly from being a suckled infant to a largely independent juvenile before they become sexually mature. Even with late weaning at 3 to 4 years of age, as is the practice in many traditional foraging societies, dependency on parents remains for some years. We all know that 4-year-olds still need plenty of parental care for housing, clothing, food support, and to survive in the community. By about 8 years of age children have just about learnt enough life skills to survive independently—they can feed themselves and seek shelter but their ability to cope with a complex and modern social situation may be somewhat limited—we recognize this as the juvenile period. This critical age for minimal life skills needed for some sort of survival is borne out by the observation that the youngest street children in our urban ghettos are about 8 years old, although even then their survival generally depends on being part of some form of pseudo-family until their teenage years.

Maturation to adulthood thus involves several elements; physical maturation, psychosexual maturation, and psychosocial maturation. At the end of adolescence the individual must be fully independent and competent across these various domains. But in evolutionary terms it makes sense for the timing of these different forms of maturation to be coordinated and thus to have coevolved so that we matched the timing of our biological maturation to that of our cognitive, psychosexual, and psychosocial maturation.

Immature social behaviour in infancy and the juvenile period is evident in all species—kittens have very different behaviours from cats, lion cubs from lions. But when they encounter their hormonal revolution they must be ready to behave as a mature cat or lion. Survival requires that in every species sexually mature animals know the social rules of their species and have the skills to seek food, and (if they are mammals) to nurture their young. The purpose of the pre-pubertal juvenile period, particularly in those species which live in social groups like hyenas and lions, is to ensure that the juveniles do indeed learn the social skills needed to be adults. Primates are no different—consider the behaviour of chimpanzees. The juvenile males and females role-play and gradually learn to be mature members of their clan. And each clan has its rituals and behaviours—the Gombe chimps and the West African chimps have to learn quite different ways of using tools to extract termites from a termite mound. Chimps of the Gombe clan poke grass stems or thin twigs stripped of side shoots into the nest, whereas the West African chimps fray the ends of sticks by chewing them to make a brush-like tool.[7] The chimps of West Africa have learnt how to crack nuts with stones and have done so for more than a century. It is a skill they learn from their parents and it takes them many years to do so, but the chimps of eastern Africa do not have this learned ability.[8] The social structures for many primates can be quite complex and have to be learnt if the individual is to survive in a clan and to reproduce.

For hominids, including humans, it would have been no different. Evolutionary pressures would have closely matched the timing of biological maturation to the timing of cognitive and social maturation. In our distant female forebears, for biological puberty to occur prior to developing the capacity to function as an adult would have been very disruptive, and probably have a high likelihood of death of the child. Even if the grandmother or siblings were able to provide additional support, the probability of infant survival would have been low. The importance of maternal maturity and experience is reflected in the greater mortality of first-born offspring even where biological and psychosocial maturation are matched. A first-born savannah baboon has only a 29 per cent chance of surviving to adulthood; a later offspring does much better but still only has a 63 per cent chance. A small study in gorillas showed that their first-born only had a 40 per cent chance of surviving the first year, but later offspring had an 80 per cent chance of doing so.[9]

Given the shorter life expectancy of our ancestral hominids, matching of reproductive and psychosocial maturation in the female would have assisted reproductive fitness. Once a female was mature enough to be a mother psychosocially it would make sense from an evolutionary point of view for her biological capacity to reproduce to be present. Either that or biological maturation must *follow* social maturation. Comparative studies do not provide any examples of reproductive competence in the female preceding behavioural maturity. The situation in the male may be different, in that in some species such as the gorilla, and perhaps early humans, the male is excluded from breeding opportunities until he achieves dominant status much later in life.

There are different perspectives on the cognitive and psychosocial attributes of archaic humans.[10] Some would argue that the skills needed to be a hunter-gatherer are not greatly different from those of an average sedentary office worker, and that a long childhood was needed to attain those skills.[11] But others think these skills did not take long to acquire. Anthropologists studying the Hazda tribe in Tanzania looked at the effects of childhood schooling on the learning of bush skills. They found that those children who did not go to school were no better hunters than those who had gone to boarding school, so childhood experience did not determine who became a good hunter.[12] Similarly studies on the Meriam people of the Torres Strait in the north of Australia suggest that the main limitation to the development of hunter-gatherer skills is simply physical—once muscular function is fully developed in mid-childhood, it is the development of size and strength at puberty, not prolonged learning, which determines efficiency in complex foraging tasks.[13]

But perhaps even more important is the development of full competence in social interactions. Abstract thought, which clearly existed by the time art, religion, and ritual appeared 50,000 to 30,000 years ago, is forming by about 7 or 8 years of age in modern humans. Older children can develop insights into social interactions and can manage some quite complex situations.[14] But the nature of society in Palaeolithic times was different, and the range of interactions needed in a clan of perhaps 25–50 people[15] was likely to be limited compared to those of a young adult in Manhattan. Thus it is reasonable to conclude that biological maturation and psychosocial maturation were synchronous and that by 10–12 years of age the earliest members of the species *Homo sapiens* were able to cope fully as adults. Thus there was a temporal match—physical and psychosocial maturation at puberty were appropriately timed. In females the first period (*menarche*) occurs late in the pubertal process and so the temporal match would have been good, because the woman by that age was psychosocially mature enough to be an adult mother in the clan.

But are these two domains—physical and psychosocial—still temporally matched in their development? Increasingly there is evidence that they are not. We would not judge the 12- or 13-year-old child, who may nonetheless have completed puberty, to be ready to face the world as a fully independent adult, capable of earning a living, building a career, rearing and managing a family, coping with a partner, and dealing with the complexities of taxes, bureaucracy, and potentially hundreds of social interactions in London or Chicago or Sydney. Clearly physical and psychosocial maturation in the teenage years are no longer matched in western society. What has happened? To answer this we need to look separately at the processes of physical and psychosocial maturation.

Our changing bodies

Many animals are seasonal breeders. This ensures that their offspring are born at a time of year—generally spring—when they are more likely to survive; it is warmer, and there is more food to support lactation. Seasonal reproduction is regulated by seasonally dependent switches in the brain controlling the secretion of reproductive hormones and permitting their secretion for only a portion of each year. These switches are sensitive to changing day length and are mediated by changing levels of the hormone melatonin.[16] For example in the male sheep the testes are shrivelled and do not make functional sperm through spring and summer. But they are reactivated and enlarge markedly in late summer and autumn in response to stimulation by the reproductive hormones. Pregnant ewes will then give birth five months later, in the spring. We can show this by tricking the ram into earlier activation of the testes by artificially shortening the day length or dosing him with melatonin, to make him think the nights are longer.

But humans have continuous rather than seasonal reproduction—when we enter puberty at the end of the juvenile period we activate our sexual hormonal systems definitively and without seasonality: it feels as if our complex and life-altering encounter with our sex hormones has begun. But actually puberty is our second major exposure to these hormones because they were transiently active during fetal life and then inactivated in infancy.[17] While a male fetus has testes making the male hormone testosterone, needed for their genitalia to develop, female fetuses have active ovaries supporting egg formation—indeed this is the only time in their lives they will make eggs, which then stay in suspended development until puberty.

Only the male fetus is exposed to testosterone and this is important for sex-specific brain development. There are structural and functional differences between the male and female brain—for example the neural control of hormone secretion is quite different, and so is psychosexual function. The size of some nuclei in the brain associated with these functions also shows gender differences. There may be other differences but this is an area where the politics of gender interact with biomedical science. The recent furore at Harvard over the comments of former President Larry Summers,[18] who implied that gender differences in brain function extended to other domains, has shown how easy it is for simple biomedical observations to be misused in a sociopolitical arena.

Even more complex is the issue of gender identity and sexual orientation—to what extent do structural and/or functional differences in brain areas established in early life influence sexual orientation as an adult? This is highly controversial and the science is still rather weak. Those who wish to show that homosexuality has a biological basis will draw one conclusion from the data while those who have an alternative view may claim that the data shows the opposite.[19] The problem is that while it has definitely been shown in the infant rat and fetal sheep that exposure to testosterone is essential for male sexual behaviour to develop,[20] the equivalent period for brain maturation in the human is much earlier (during fetal development) and the data are very obscure and not compelling. Thus although there are gender-based differences in some brain functions related to reproduction, the evidence is far from conclusive that early life hormonal patterns relate to gender identity.

Soon after birth the initial activation of the sex hormone system is curtailed by mechanisms within the brain. It stays quiescent until that most critical of hormonal revolutions: puberty. At that time the brain reactivates the system which releases reproductive hormones from the pituitary gland. This leads to the testis secreting testosterone and developing the capacity to make mature sperm, and to the ovary starting to secrete oestrogen. In the male, the testosterone causes pubic, facial, and axillary (armpit) hair to develop, the larynx (voice-box) to change shape and shift its position downwards in the neck leading to the voice deepening, and skeletal and muscle growth. This is demonstrated from the lives of the 'castrati'—Italian eunuchs, castrated before puberty to ensure that they maintained a high singing voice for services in the Roman Catholic Church. The last castrato of the Sistine Chapel, Alessandro Morereschi, died in 1922. His is the only castrato voice to have been recorded for posterity—it shows a quite distinct and unusual timbre. In girls the sex hormones, particularly oestrogen, cause the breasts to grow, pubic hair to develop, and promote the growth spurt.

The entire process of biological maturation takes several years. The first sign of puberty in girls is breast development and in boys testicular enlargement. In both sexes this is accompanied by an acceleration in height. From this point to the end of puberty takes three to five years. Puberty is over when the individual is fully grown—and this is when the influence of the sex hormones has led to fusion of the growth plates in the long bones and so growth is no longer possible. After that time there will be still some further muscle development, and certainly in modern society there will be further brain development, but the individual is now adult-sized and is reproductively competent.

But there are big sex differences in the time at which reproductive competence occurs during the pubertal process. Boys can produce sperm from relatively early in puberty—much of their pubertal growth spurt follows the development of their biological capacity to reproduce. Girls on the other hand have virtually stopped growing before their first menstruation, which in turn reflects the hormonal changes following their first ovulation. This occurs late in puberty as a result of the brain's pituitary hormone release changing to a pattern that triggers maturation of an egg within the ovary, followed by its expulsion into the Fallopian tube. A process of cyclic hormonal change is now established which leads to the uterine lining growing then being shed (which creates the menstrual flow) on roughly a four-week cycle.[21] But many of the initial cycles are associated with failure of ovulation of a mature egg and so the periods may be irregular for a year or so and fertility is low for about one to two years before full reproductive competence is reached.

Humans are different from other mammals, including other primates, in having a growth spurt in skeletal size during puberty. Most other animals terminate their growth at the time they enter puberty. Why is there this difference? It is an intriguing question about which there are a variety of views. One is that it arose because humans had to delay their growth because they needed to invest the available energy in brain growth and thus delayed skeletal growth until brain growth was complete.[22] An alternative view, which we favour, suggests that in males the pubertal growth spurt arose through the process of sexual selection; the taller male either being more attractive or more dominant in a mating hierarchy.

In the female natural selection must have played an important role in the evolution of the growth spurt because of the critical importance of pelvic size. Pelvic diameter is directly correlated with height and it is no accident that it is not at its maximum until late in female puberty, at about the time of attainment of full fertility—one or two years after menarche. Short stature therefore leads to a smaller baby through the processes of maternal constraint, which limits food supply to the fetus, and smaller babies fare less well. So natural selection will have operated to favour females who grew a bigger pelvis by the time that they were reproductively competent.

Adjusting the clock

There are genetic factors which influence the timing of puberty. The tendency to early or late puberty runs in families, and some populations do indeed have later or earlier menarche—northern Europeans for example tend to have later menarche than southern Europeans by an average of about seven months.[23] But more importantly there are a number of ways in which the environment can influence the timing of puberty. Of these nutrition is critical but the effect of nutrition differs before and after birth. While poor childhood nutrition *delays* puberty, poor fetal growth may lead to *earlier* puberty.[24]

If fetal growth is impaired by deficient nutrient supplies the fetus may predict that its future is going to be bad, and it thus chooses to adopt a life-course strategy more like that of the smaller mammals—that is to accelerate sexual maturation to ensure gene transmission to the next generation. There is evidence that life-history biology applies to humans as much as it does to insects. The effect of low birth weight to accelerate puberty is generally small, perhaps advancing it by only a couple of months. However the effects of postnatal nutrition are larger, so poor fetal nutrition combined with good postnatal nutrition can advance puberty by well over a year.[25] In that situation the fetus has sensed that it will be born into a dangerous world and adopts the strategy of trying to accelerate maturation; but it can only do so, at least in the female, if there is sufficiently good nutrition in adolescence for her to be able to support a pregnancy. This is a situation where the phenotypic manifestation of the fetal environment is dependent on the postnatal environment. This interaction is most dramatically observed in children who were born in very poor societies but then adopted and brought up in the richest countries—their rapid switch from poor early nutrition to good childhood nutrition is associated with much earlier puberty—with some girls having their first period as early as 6 to 8 years of age.[26]

In contrast poor nutrition and ill health in childhood delays puberty. Some human biologists argue that this is a form of biological trade-off, which enables more pre-pubertal growth to occur before reproductive maturity. This may have been critical in the female to ensure that she did not get pregnant until she was in an adequate metabolic state able to support the nutritional demands of both herself and her fetus. This delaying response appears to be a general strategy for slowly reproducing species such as humans—i.e. to postpone puberty in poor conditions in the hope that in time conditions will improve.

Thus the timing of puberty seems to be influenced by at least three things: genetic factors, the early developmental (prenatal) environment, and the later nutritional environment of childhood. There are close parallels between this and the discussion that will follow in the next chapter on the factors determining metabolic regulation. This is not surprising. Growth and metabolism are two key elements of every organism's strategy to optimize its fitness and maximize the chance of its genes being transmitted to the next generation, and these processes are intimately linked. Indeed the brain mechanisms used to regulate the timing of the onset of puberty overlap with those regulating metabolism—so both at the level of the organism and at the level of brain wiring, food and sex are linked.

The timing of puberty may also be confounded by other factors. One that has received much recent attention concerns the levels of chemicals found in the environment which interact with the sex hormone control mechanisms. Such chemicals abound in modern foods—whether they are naturally occurring as in soya products, the widely used artificial growth promoters found in meats, plasticizers in bottles, or the metabolites of oral contraceptives and hormone replacement therapy passing from urine into the water supply. While these so-called 'endocrine disruptors' have been implicated in changing patterns of hormonally related tumours such as breast cancer, the evidence that they have a role in altering the timing of the onset of puberty in modern humans is far less compelling.[27] But it is an area that needs much more research and one which has to be monitored carefully.

Some data also suggest that various forms of stress can either advance or delay puberty by affecting the functional settings of the complex brain networks controlling hormone release and again these effects may depend on when in development the individual experiences stress. Thus there could be spill-over into the systems controlling the onset of puberty.[28]

Palaeolithic puberty

Why did we evolve the capacity for the onset of puberty to be triggered possibly at around 7 to 8 years in girls? The answer lies in our hominid pasts and we have to rely on the detective work of archaeologists and anthropologists to infer the characteristics of life as it was more than 10,000 years ago.

The fossil and skeletal evidence is obviously limited in terms of what can be known for certain but considerable insight can nonetheless be gained, particularly where large numbers of skeletons can be examined. Other pieces of information can be gleaned from modern hunter-gatherer societies but, in contrast to those of 10,000 years ago, modern hunter-gatherers often live in environments on the margins of colonization. These modern societies tend to be somewhat deprived, at-risk populations, whereas Palaeolithic hunter-gatherers lived in environments they selected for the adequacy of food supplies and the security they provided. Fossil evidence does not suggest malnutrition to be a major feature of human Palaeolithic lives—that came later with the development of farming and towns.

But life expectancy was short in the Palaeolithic period. If one survived childhood, and the chance of doing that might be only 50–60 per cent, then living beyond middle age was still unlikely—and old age would be unusual. We can do some simple calculations of the timing of puberty, based on our best evidence about the variables which need to be taken into account—namely the chance of living to adulthood, of dying in childbirth, and the need for sufficient live children to survive to reproduce for the population size to remain stable. We can factor in the spacing of the babies women would have delivered—assuming breast feeding behaviour comparable to that of modern hunter-gatherers and/or the practice of infanticide to ensure child spacing. We must also allow for the fact that children need their mother's survival to an age when the youngest child required to ensure population stability has reached reproductive maturity.

The formula makes some assumptions, and by far the biggest influence on the calculation is the estimate of life expectancy. But if we do the arithmetic, based on a generally accepted average Palaeolithic life expectancy of 35 years for a female who survived infancy and childhood, an average of two children surviving to adulthood (including at least one female) so that the population could remain stable, and allowing the mother to live long enough to support her youngest survivor to a fully independent existence, we find that she requires a reproductive span of 16–18 years. That means there would be advantage in being reproductively competent by 13–15, and this means having menarche at 11–13 and breast development starting perhaps at around 7–8 years of age. We think that this is the underlying pattern of growth and maturation which evolved for our species.

Maturing minds

In both sexes the physical changes at puberty lead to the recognition that the individual is under the influence of his or her hormones and parental and social attitudes to the boy or girl start to change. The hormonal influences on the brain drive the development of psychosexual function—the development of concepts of self and views of one's sexual identity emerge during puberty. But psychosexual maturation cannot be separated from reproductive maturation—they are co-dependent. Without a rise in sex hormones at puberty psychosexual maturation will not occur, but hormonal changes alone are not sufficient—the brain must be mature enough to respond. Children with extreme abnormalities of development, for example due to brain tumours, may undergo precocious puberty which gives them the physical sexual characteristics of a much older individual, but they do not become psychosexually mature until much later.

But there are also significant changes in other components of brain function. Many aspects of brain development such as cognitive maturation have timetables independent of the sex hormones. That is why in high schools there is no difference in examination performance between those 13-year-olds who have completed puberty and those who have not. During adolescence there is a burst of rewiring of brain circuitry.[29] Indeed recent studies using sophisticated imaging techniques show that the brain has its greatest connective capacity in early puberty. But we are also learning from these same studies that some changes in brain function and wiring continue until much later and that these particularly involve the connections from the prefrontal cortex. This brain region is the last to mature and is involved in the development of attributes such as responsibility and self-control—it is generally thought that the sometimes risky exploratory behaviour of early adolescence reflects this immaturity of the prefrontal connections. We are older but wiser when these late-maturing systems are fully active. Perhaps the rental car agencies in New Zealand are right in not allowing rentals to people under the age of 25?

Unfortunately we can never be certain from anatomical studies whether it now takes longer for the brain to be fully mature in modern society, because the necessary imaging techniques were not possible even five years ago. Anecdotal evidence would suggest that it does. Two centuries or more ago many teenagers could, and did, take on roles as mature adults. Midshipmen (junior officers) in the Royal Navy during the Napoleonic Wars were in their early teens—the future Lord Nelson joined his first ship at the age of 12. Our own grandfathers and great-uncles travelled the world independently of their parents to build a new life and they started businesses in their mid-teens. Has the cycle of building brain connections as we learn, then refining and pruning these connections, changed? Or do adolescents need to know so much more to become an adult in a complex society? Or is it that as society has got more complex, we treat adolescents differently and inhibit their maturation? Might external influences like the media and loss of tight societal pressures have reinforced exploratory behaviour and altered the development of behaviourally inhibitory pathways? It will be hard to answer these and many other similar questions. Clearly adolescence is a changing and complex psychodrama in action: it involves a cast of internal and external characters. Our self-image, the image that others have of us, and the functions of our brain all undergo dramatic changes as we move from juvenile to adolescent. Our capacity for abstract thought develops in late childhood but our brain connectivity continues to mature. Cognitive and social 'intelligence' develop in the context of the society we live in, and are related to what society provides and requires of us as individuals.

Changing times

So while we evolved 150,000 years ago with synchronous maturation of our bodies and our brains at puberty, there have been significant changes in the timing of both over the last 10,000 years. These can be understood in terms of the major transitions we have faced. For quite separate reasons maturation of both body and brain were delayed by environmental factors until recently and therefore stayed in synchrony. But over the last hundred years they have diverged; while psychosocial requirements have become more demanding and full maturation appears to have been delayed, physical maturation is getting earlier. A temporal mismatch has been created. Why has this happened and what are the consequences? This is the topic of the remainder of this chapter.

The development of agriculture brought settlement, and settlement brought concepts of property and the development of a new social structure. Agriculture brought humans into proximity with animals and into greater proximity with each other as populations grew—static settlements dependent on agriculture allowed more people to live at one place. The risks of diseases from cross-infection were thus increased. In addition there were fundamental changes in our nutrition. While hunter-gatherers had multiple ways to obtain food, populations dependent on a fixed location for their herd and crops became inevitably more at the mercy of climate and war. Malnutrition and infection affect children first and their growth was reduced. When childhood nutrition is poor, puberty becomes delayed and so with changing patterns of settlement came a delay in puberty.

But at the same time social structures became more complex, and the skills needed to thrive in society would have increased and probably taken longer to learn. Even though changes in physical development on one hand, and psychosocial development on the other, were not directly linked, the net effect would have been roughly parallel, with both being progressively delayed. The underlying drivers for delayed biological puberty and for greater psychosocial skills as an adult were ultimately linked by a common element—increased population density.

Larger populations also stratify tasks—there would be specialist toolmakers, bakers, and soldiers. Settlement also created property rights, and property rights bring hierarchies, and custom and law and taxes and power structure, and these bring stratification of society—from the rich man in his castle to the poor man at his gate. The history of Europe from 2000 BCE to the eighteenth century is one of increased population density, punctuated by episodic declines during epidemics and wars, and increasing social complexity. It culminated in hierarchical feudal and monarchal systems. But from the perspective of the age of puberty, these social changes through to the eighteenth century did not matter—the time to achieve status as an adult may have been longer but so was the time of physical maturity. They were still matched. Indeed if physical maturation occurred after psychosocial maturation for the female then the match had a fail safe aspect—women would not be placed in a position of reproduction until ready for it.

Thus while the timing of both physical and psychosocial maturation changed we believe that the necessary relationship between them persisted from when we first evolved until very recently. But suddenly, at least for European populations, it has all changed.

The study of growth (sometimes called auxology) started in France in the late eighteenth century when Count Philibert Gueneau du Montbeillard measured the growth of his son over eighteen years from 1759 to 1777 and kept accurate records, just as so many parents do today using marks on the door frame. He was the first to record that human growth could be episodic and to document the pubertal growth spurt. By the mid-twentieth century the growth rates of children and the final heights they achieved were being used to monitor public health.[30] While there might be genetic differences between populations—pygmies from Zaire are short, the Watusi from Tanzania are very tall, it was *within* population trends that economists and public health officials particularly examined. It was noted that western populations were getting taller and taller and that menarche was occurring earlier and earlier. This is called a 'secular trend'. Historical auxology became a discipline and old records were examined to find out when such secular trends had started.

Studies of the timing of menarche across western populations in the last 150 years have shown a dramatic decline in the age at which it occurs—generally about three months' decline in the age of first menstruation per decade.[31] The process is remarkably parallel in different countries, albeit starting from different points which may in turn reflect underlying genetic differences between populations. We must not however assume that the European experience is universal. Many populations in the developing world never showed menarche occurring as late as it did in Europe.[32] This may reflect genetic differences, or it may reflect the generally better conditions of pastoralist or forager societies compared to the squalor of serfdom and urban slums in pre-Enlightenment Europe. Neither would they have confronted the complexities of large population densities, at least until colonization.

We are led to the conclusion that this secular trend in the age of menarche is the consequence of the great improvement in health, nutrition, and maternal care in Europe. From the Enlightenment at the end of the eighteenth century came a greater focus on the plight of the child. In England social reform movements appeared and by the mid-nineteenth century they were in full swing. Child labour laws started to be enacted, nutritional support and charity for the poorest members of society began to appear. But perhaps most importantly public health initiatives appeared. The open sewers of the worst cities started to disappear. Concepts of hygiene and public health arose and slowly this impacted on both the health and the life expectancy of children.

Studies of the age of menarche show that the trend is now levelling off in those countries with the longest history of good child health. This supports our conclusion that the age of menarche is now returning to a timetable set by evolutionary processes—i.e. approaching its age selected by evolution as the constraints of undernutrition and poor child health are removed.

But what about the society young people now live in? It is clearly much more complex. The time needed to be fully functional as an adult has increased markedly, indeed many young people appear not to be able to stand alone in a twenty-first-century urban jungle until their twenties. For the first time in our evolutionary history our psychosocial maturation occurs *after* our physical maturation. But at what cost?[33]

Mismatched maturation

This maturational mismatch is a dominant issue for western society. *Time* magazine puts it on its cover. While early puberty is blamed on exogenous factors such as hormonal contaminants of foods etc. we do not think they are the heart of this mismatch (although they may be very important to other issues such as the changing incidence of breast cancer)—because the trends of falling age of menarche started as social conditions started to improve over 100 years ago, and the trend towards a progressively more complex society started thousands of years ago.

So there is a growing problem—youngsters are biologically mismatched to the society they live in, which was designed around the expectation of adulthood appearing almost a decade later. And, rather than confronting its origins, when the problem is discussed it is often medicalized—if girls enter puberty too soon perhaps hormonal treatments should be offered to delay puberty. But is this correct? We would argue that these children are simply revealing their evolutionary origins—the constraints on their development have been largely removed and they are entering puberty at an age little different from their Stone Age forebears.

For these reasons we are concerned about the use of the term 'precocious puberty' being applied to normal children having earlier puberty. Precocious puberty is strictly a medical term to describe those with organic disease which leads to pathologically early puberty[34]—for example some brain tumours can lead to puberty being seen at 2–3 years of age. To use the term 'precocious' implies abnormality, when what has happened is that the timing of puberty has returned towards its genetically determined point, albeit with consequences of the mismatch.

The mismatch of modern puberty is a fundamental and irreversible issue. For the first time in our 150,000-year history as a species, the norm at least in western society is now to achieve physical maturity well before psychosocial maturity. Children are not (we hope) going to get less healthy and so they will continue to enter puberty early—and as global health improves, more and more children around the world will enter biological puberty between 7 and 10 years of age and be reproductively competent from a biological perspective by the age of 11 or 12. On the other hand society is not going to get less complex, and the skills needed to be a successful adult are likely to increase still further. We will need our increasingly long childhood to acquire these skills.

In developed societies we expect a good deal of our young people—it is an expectation driven by social emancipation and by the impact of media messages. As a society we confuse physical maturation with psychosocial maturation so we have a tendency to assume a person who is biologically mature is a full adult, and vice versa. In our growth clinics we see this so often: a person with a growth problem leading to a severely short stature may be an adult but is treated by society as a child; a 6-year-old with true precocious puberty due to a disease may look like a 12-year-old and be treated that way—impossible. This transference of impression and expectation increases pressure on all our children, not just those with truly precocious puberty. It is magnified by the media, the entertainment and advertising industries, which put impossible pressures on these physically mature but psychosocially immature adolescents. Indeed, the media and marketing industries play on this mismatch and exploit the latent sexuality of the late pre-teens and younger teenager. Teen magazines, popular music videos, and television programmes for this age group are increasingly dominated by latent or even overt sexuality. This further complicates matters, by driving the psychosexual expectations of these young people while their other psychosocial skills remain underdeveloped for many years to come.

Many of the problems of our young people are exaggerated by this mismatch. Think of a class of 15-year-olds in a school today, many of whom will be physically mature, if not sexually active. Their acting out and exploratory behaviours represent their attempts to live in a society which expects them to function as adults just because they look physically like adults. They cannot do so consistently because their psychosocial maturation is not yet complete—there are so many aspects of how society works which they cannot get in perspective. Society hedges their behaviour around with rules and regulations, dictating that they cannot be responsible to drive or vote until they are older—to their even greater frustration.

And this has all happened very fast—their great-grandparents and even their grandparents did not have to confront this issue—so parents and grandparents, the generations in control, still perceive adolescence in relation to their own pubescent experiences which were very different. Indeed many of the structures of society which today affect young people, such as the organization of middle and high schools, the mores to which we expect young people to conform, our institutional attitudes to sex education, etc., were largely established in Victorian times and have undergone surprisingly little change since then.

But there is a further way in which the experience and attitudes of previous generations may make them uncomfortable or ill-equipped to advise young people on these issues. For most of human history, sexual activity was almost inevitably linked with reproduction and there were good societal, economic, and biological reasons why societies developed an orderly procession of roles in life. In European societies this evolved into a sequence of education, career (for men, anyway), courtship, marriage, sex, children. Social convention, sometimes disguised as moral guidance but with a basis in the realities of primate mating strategies, dictated that solid (and preferably legally or societally enforceable) relationships should be established before reproductive activity was sanctioned. But in modern western society the easy availability of reliable contraception means that sexual activity is clearly distinct from reproduction. This separation may have encouraged the commercialization of sex—the messy details of human birth don't sell many cars—but it has also changed the behaviour of young people, who are now able to use sexual intercourse to begin and explore relationships with much less regard for the biological consequences. We are seeing rapid cultural evolution driven by a technology-induced change in behaviour – and a discordance has developed between the attitudes of the young in our society and the social structures established over previous generations.

Sex education seems to be based on the assumption that young people do not 'need to know' until well into their teenage years. But is that any longer realistic? Children as young as 7 and 8 must confront overt pubescent changes in their bodies and they need to understand what these changes mean. These pre-adolescents can be reproductively competent by the age of 10 or 11. Is it proper or sensible that appropriate information about their sexuality is not provided? Our own attitudes to adolescent sexuality are driven by the prior assumption that biological and psychosocial maturation are coincident. They are not now and are unlikely to be so in the future. Biological maturation for many will precede psychosocial maturation by up to a decade. Just saying 'they don't need to know' and that they must practise abstinence appears increasingly unrealistic in biological let alone psychosocial terms. These are very hard issues to grasp for those of us brought up in a different time, whether of liberal tradition or not. They cause us much internal conflict, and they pose hard questions to address.

But we need to find the answers—how should we help young people to manage their biological maturation ahead of full social maturation? The more we think about this mismatch the more challenging it is for us and our society. Do we have to change the way we educate children and adolescents? Do we have to adjust our mores and attitudes to the reality of earlier biological maturation? Can we learn from other societies that have evolved with apparently much less adolescent turmoil? Can we find new approaches to redress some of the pressures on young people in an age where the information explosion and potential for exploitation is so much greater and controls on their behaviour are so much more unrealistic? We do not know the answers but what we are sure of is that ignoring the growing mismatch of puberty and not recognizing its fundamental and irreversible nature is harming our young people and our society.

7
A Life of Luxury

The town of Albi sits on a bend in the river Tarn in southern France and is dominated by an enormous Gothic cathedral. The Bishop of Albi and vice-inquisitor Bernard de Castanet started building the cathedral in 1281 after the conclusion of the Albigensian Crusade by the Roman Catholic Church against the Cathar sect[1]—the sumptuous internal decoration is intended to make a political and religious point of contrast against the extreme asceticism of the Cathars. This cathedral is in a sense a memorial to the terrible conflict over religious belief that has been a feature of the lives of *Homo sapiens* for at least the last 2,000 years. But next to the cathedral lies an extraordinary museum. It has two sections that might initially seem to be totally discordant—but on reflection they suggest much about our evolution as a species. The popular attraction is the Toulouse Lautrec gallery—for this great documenter of hedonistic Parisian society of the late nineteenth century was born in Albi. The magnificent paintings and posters of the unhealthy lifestyle—of the dancers, the poseurs, the pimps and prostitutes, usually in settings of smoky bars—were nonetheless drawn or painted with pathos. But on the second floor of the museum is the regional archaeological collection. One of the smallest items on display, easy to miss in a cabinet with many other artefacts, is a small figurine carved some 25,000 years ago by an unknown genius, for this region is where the magnificent artistic expression of early *Homo sapiens* is first demonstrated in its full glory—the caves of Lascaux with their exquisite drawings of animals from 17,000 years ago are about 160 km away.

Venus figurines carved from ivory have been found at a number of sites across south-west Europe.[2] They were carved between 20,000 and 30,000 years ago and are some of the earliest pieces of representational art known. They typically depict a buxom, well-rounded (or frankly obese) female. We can only guess their purpose and significance. It is generally assumed that they are fertility symbols that represent bounty, fertility, and health.

Are they memorials to hope of a more productive kind than the circumstances represented by the enormous stark cathedral, or Lautrec's paintings? To us it seems ironic that some of our earliest representational art may reflect the beauty of an obese figure yet obesity is arguably now our biggest health concern.

As a result of the major changes in our environment there are now more people across the globe suffering from the effects of being overweight or obese than from undernutrition despite the dire poverty of sub-Saharan Africa and parts of Asia. In the last few decades there has been a massive rise in many countries in the incidence of so-called 'lifestyle' diseases and in particular cardiovascular disease, obesity, and adult onset diabetes. In part this increase is exposed by our longer lifespans, because these have been generally considered diseases of middle and old age. But now we see a rapid rise in the levels of obesity even in children as young as 3 years, and the appearance of one of its major consequences, so-called 'adult onset' (non-insulin-dependent or Type 2) diabetes in the second and third decades of life.

In large part these changes are ascribed to a new kind of lifestyle: less exercise and more high-energy food. The average weight of North Americans increased by 4.5 kg during the 1990s—it is estimated this cost the airlines an additional 1,600 million litres of jet fuel to fly them around during 2000.[3] A new class of lawsuits has arisen where airlines have been sued because they insist that an oversize passenger must buy two seats.[4] Clothes shops now order a very different size range of clothing than they did a decade ago. Even the standards for toilet seats are being revised to cope with heavier people.[5] In western countries well over 20 per cent of adults are now classed as obese. The prevalence of overweight children rose from 8 per cent to 20 per cent in the UK between 1984 and 1998. And even in a country undergoing economic transition such as India, 10 per cent of urban middle-class children can be classed as obese or overweight.[6]

Why should these changes in lifestyle lead to an increased disease risk? After all humans are a generalist species who can live in almost any environment. Why should living in a rich environment lead to increased risk of disease?

The human engine

Nutrition is much more than just type and amount of food consumed—it is about the balance between fuel supply and expenditure. It is this balance which creates our real energy crisis. Our human engine runs on energy, and that energy ultimately comes from food. The major fuels are simple carbohydrates. After digestion complex carbohydrates, fat, and proteins can be turned into energy sources by conversion to glucose. This largely happens in the liver and skeletal muscle. Without fuel no cell can function. After glucose is taken into a cell, it is converted to energy through the action of mitochondria,[7] which are intracellular incinerators generating energy for use in an enormous range of body processes. We use energy to run all our cellular processes as well as to carry out integrated actions such as muscular movements. The electrical processes in the brain are particularly energy consuming. And a further important use of energy is in the repair and maintenance of our tissues—for example the cells of our skin and of our intestinal lining need constantly to be renewed.

Growth is not possible without the adequate intake of fat, protein, and carbohydrates—the so-called macronutrients, which provide the building blocks for tissue growth (e.g. muscle, bones) as well as the energy that allows cells to divide and multiply. But good nutrition for function and growth must also include micronutrients such as the vitamins and trace minerals. These are essential for specific body functions, often as the catalysts for specific enzyme actions or as components of critical molecules. For example iodine is essential in the body as a component of thyroid hormone.

But fuel supplies must be balanced by their consumption. If people expend less energy than they take in they will gain weight. When excess energy is left on board it is primarily stored as fat under our skin and within our abdomen (so-called visceral fat). If we have a persistently excessive energy intake, fat will also accumulate in our muscles and in the liver. Fat stores are a long-term energy supply, just like the camel's hump—an adaptation in that species which evolved for surviving in an environment where there is very intermittent access to food (the hump may also act as a thermal insulator).

Fat has the highest energy content for its weight of any body constituent and that is why animals deposit fat under particular situations where having fuel reserves is important. For example a hibernating animal will have a very high body fat content when it enters hibernation and low levels when it wakes from its winter sleep. Even more dramatic are migrating birds—in July the bar-tailed godwit gorges on clams plucked from the inter-tidal mud in the estuaries of the Alaskan peninsula. It eats and eats until its body fat forms in thick rolls. At the same time its liver, kidneys, and intestines shrink dramatically. Fat will form over 55 per cent of its body weight at the start of its 11,000 km migration across the Pacific Ocean to New Zealand. These birds fly in large numbers at 70 km per hour without stopping—a journey that lasts four or five days. They do not eat during the journey, relying totally on their body stores. The fuel supplies are totally exhausted by the time they arrive in New Zealand—having completed the longest non-stop migration of any bird.[8]

There have been considerable reductions in our personal energy expenditure since the industrial revolution, although this has been accompanied by a massive increase in the other form of energy consumption through use of electricity and transport fuels. In the developed world the burden of physical labour in the course of work is greatly reduced by machinery. There is a dramatic correlation between motor vehicle ownership and adult obesity in China.[9] In many parts of the developing world women and children still expend much energy in order to collect water, firewood, and to do the washing, but these are trivial exercises in the developed world. In India, China, and Thailand the rise in childhood obesity can be correlated with a reduction in walking, cycling, and other forms of exercise. Even in our leisure we have replaced physical activity through sport with television or sedentary electronic games—the time spent watching television as a child or adolescent predicts markers of poor health such as overweight, physical unfitness, and blood cholesterol level as adults.[10]

The control of body weight is complex. People have different body weights, not only because they are taller or shorter but because they have stored different amounts of fat. And in the absence of significant changes in habit, our body fat content remains relatively constant—winter or summer, at work or on vacation. Thus part of the variation in body composition between individuals appears to be innate; consequently it must have something to do with our individual underlying physiology. It cannot simply be due to differences in supply or demand. It is true that some obese people overeat, but equally it is true that some do not; they maintain a higher weight even on a diet that might make others lose weight. Similarly, some of us manage to stay slim on a diet that will make others gain weight fast. The differences lie in the complex control systems our bodies use to regulate fuel supply and consumption, and as we will see the major settings of these control systems are established early in life.

We have seen how species evolved to primarily live within their comfort zones. Our energy system—that is, the nature of our metabolism including the fuel it burns, how it burns it, and what it does with excess fuel, are all the products of our evolutionary history and therefore of the environment we inhabited in prehistory. But our metabolism was designed for environments very different from those we now inhabit. We need to examine the consequences of this change.

Thrifty bodies

The idea that there might be a mismatch between our metabolic system and the environments we live in was first proposed some three decades ago by the geneticist Neel who suggested that genes had been selected in our evolution to help us to survive as hunter-gatherers.[11] He argued, probably mistakenly, that hunter-gatherers experienced feast or famine and that while the former posed no threat humans were selected with a repertoire of 'thrifty genes' to enable them to survive famine. Such genes affect a range of metabolic processes including those designed to promote fat storage. When humans live in a modern world of plenty, these same genes continually drive fat deposition and so we suffer the consequences in terms of diseases such as obesity and diabetes.

We now think that Neel's original premise was incorrect, because hunter-gatherers had relatively good nutrition, were able to move to follow food supplies, and experienced relatively little famine compared to what followed with the development of agriculture.[12] Neel's idea—the 'thrifty genotype' theory—led to a large industry of geneticists looking unsuccessfully for such thrifty genes. But perhaps Neel had the story partially right, because genetic variation probably does play some role in the origin of diabetes, although it does so much more indirectly, because it changes the sensitivity of the individual to his or her environment.[13]

An initially surprising set of observations has shifted our attention towards the role of environment during development. The first clues came some fifteen years ago when an unexpected association was uncovered from a series of population-based studies—these showed that the smaller a baby at birth, the higher the risk of its dying of heart disease or of developing diabetes in middle and old age. The story of this discovery, made by our colleague David Barker from Southampton, is described in *The Fetal Matrix*. Initially it was met by disbelief among scientists including other epidemiologists. How could it be that events happening before birth change one's risk of developing heart disease or diabetes fifty or so years later? Many experts from a wide range of disciplines have tried to refute such findings, but they have stood the test of time and have now been confirmed by many other studies.[14]

In the last ten years we and others have shown that we could mimic these observations in experimental animals and made a number of physiological observations which gave a solid scientific basis to the association.[15] Most recently it has been shown that at least some of this biology is underpinned by epigenetic modification of specific genes.[16]

These effects need not be unidirectional. Although we focus especially on the more frequent scenario of moving from a more limited environment early in life to a richer environment later, there are data which suggest that switching from a rich environment *in utero* to a poorer environment after birth can have consequences. In famine conditions in Ethiopia it is the babies that are born larger who have a far greater risk of developing rickets, a disease of bone which is due to inadequate amounts of vitamin D.[17] And although the data are limited they suggest that big babies who then face poorer conditions are at greater risk of developing diabetes.[18]

These epidemiological, clinical, and experimental observations are part of a bigger story about how developmental signals constrain and determine the range of environments we can live in healthily. If we live outside that zone then we are mismatched to that environment and we are more likely to get disease. In turn this has led us to develop concepts about how developmental processes operate and how they respond to the environment. This science, sometimes called 'ecological developmental biology', is young and until now has largely been the domain of those interested in plants, insects, amphibia, and reptiles. We now see how it applies to human biology.

Within the womb

The concept that the fetus responds to environmental signals is both very old and very new. Hippocrates realized that the health of the fetus was dependent on the health of the mother.[19] Indeed, the work of embryologists such as Spemann and Stockard in the early twentieth century revealed critical periods of sensitivity to both internal and external environmental stimuli during embryonic development.[20] But it was not until much later that doctors really started thinking seriously about how the fetus was affected by factors impinging on it from the world outside. In the 1940s, Norman Gregg, an ophthalmologist studying the origin of congenital cataracts, came to the tragic recognition that rubella (German measles) infection during pregnancy could cause birth defects. In 1961, another Australian, William McBride,[21] recognized that thalidomide, a sedative prescribed in pregnancy, could cause limb defects. In the 1960s, 1970s, and 1980s there was an explosion of knowledge about the fetus due to the development of experimental techniques for studying animal fetuses, particularly sheep fetuses in the womb. We learnt that there are many other ways in which the fetus is affected by the external world. If the mother got a fever, so did the fetus; if the mother was stressed then some hormonal signals of stress, albeit dampened, crossed the placenta from mother to fetus. If the mother's nutrition changed significantly, the fetus would detect changes in nutrient delivery across the placenta. If the mother had low blood oxygen levels due to disease or being at altitude, the fetal oxygen level also fell—and many more examples were found.

Against this background, several lines of research have taught us much about how fetal growth was regulated. It has some very different features from growth beyond infancy. Adult height is strongly influenced by genetic factors—that is why there is a strong statistical relationship between parents' and their children's adult heights. But the genetic determinants of fetal growth are much weaker and the mechanisms of growth control very different.[22] The size of the fetus is dependent on features of the maternally created environment largely independently of genetic effects, so as to ensure that the fetus can pass through the pelvic canal. In a classic experiment from the 1930s, Walton and Hammond crossed Shetland ponies with much larger Shire horses and showed that it was maternal size which was the primary determinant of the size of the foal at birth.[23] More recently, with the development of assisted reproductive technologies, these experiments have been repeated in a more elegant way and the same conclusion reached—namely that it is the environment created by the mother rather than her genes—or indeed the fetal genes—which has the major influence on fetal growth.

Indeed studies in humans where eggs have been donated show that birth size is much more closely related to the recipient's size than to that of the donor.[24]

Fetal growth is ultimately dependent on how well the mother can supply nutrients and oxygen across the placenta. There is no other way that the fetus can get food and oxygen. The placenta must also remove waste. We know a fair amount about how this complex supply line works in the last two-thirds of pregnancy when the placenta is fully functional. We also know that there are many ways it can be disrupted. First, if maternal health is compromised then available nutrients will be prioritized to assist her and the fetus will receive less. In evolutionary terms the fetus is expendable because if the mother lives to have further pregnancies then her gene transmission has been preserved. If however the mother dies, then so does the fetus and there is no passage of genes to the next generation. Secondly, the supply line is dependent on how well blood is pumped to the uterus. This can be interfered with by disease. Thirdly, the placenta must function properly, but in doing so it will consume some of the oxygen and nutrients provided by the mother. The placenta can be affected by infections such as malaria and its blood vessels damaged by maternal diabetes and pre-eclampsia (a disease which affects the blood vessels of the placenta). Only a fraction of the nutrients consumed by the mother get to the growing fetal tissues, and then only if the fetal heart is pumping well and if the fetal hormonal state is favourable.

Thus what the fetus gets as nutrients is not simply what the mother eats. She eats to meet both her own needs and those of the growing fetus but their nutritional requirements are different. The placenta secretes hormones into the mother to change her metabolic regulation, making it less dependent on glucose, and this allows more of this fuel to be provided for the fetus. Meanwhile the mother's metabolism becomes more dependent on other nutrients such as fats. We now know that under famine conditions, when the maternal food intake is below 800 kcal per day in mid- and late pregnancy, birth weight is grossly reduced.

But even much more subtle changes in maternal diet can induce changes in fetal growth and development.[25] In developing countries both poorer maternal nutrition and high workloads are associated with lower birth size.[26] Indeed in developed countries those who undertake severe exercise in pregnancy such as running marathons give birth to smaller, leaner babies.[27]

However in general the consensus is that moderate exercise provided it is balanced by appropriate nutritional intakes is safe. But research now suggests that even changes in the balance of the maternal diet which do not alter birth size can nonetheless have echoes on the child for the rest of its life.[28]

In both animals and humans, many of the most important changes in gene expression occur during the first weeks of pregnancy, including the first six to eight weeks when the woman may not even be aware that she is pregnant. This is the time, particularly in the first week, when epigenetic changes are occurring to the DNA determining the pattern of gene expression which will not only control the next stages of its development but also some of the fetus's attributes throughout life. It is a time of enormous interaction between the fetal genes and the environment. Studies show that as well as nutrition in pregnancy having an influence on birth size, maternal nutritional state at conception is important.[29] This suggests that it is the long-term nutritional state of the mother which determines how nutrients are mobilized to support fetal development. This creates a major public health issue—how can we optimize pregnancy outcomes when some of the major factors involved include the time before conception? Less than 50 per cent of women actively plan when they will conceive. The challenge may not only be limited to the female; there are some data beginning to emerge to suggest that environmental factors mediate epigenetic changes in sperm which may also play some role in affecting prenatal development and the later health of the offspring.[30]

We have focused on the provision of the major nutrients glucose and oxygen because they dominate the regulation of fetal growth. But the fetus must get the full repertoire of fats, amino acids, and micronutrients including vitamins from the mother. Our studies in the Sherpa showed the consequences to the fetus of iodine deficiency. If the mother is micronutrient deficient so will be the fetus. This is why it is strongly advised that diet in early pregnancy is supplemented with iron and folate. In many parts of the world there is a much broader concern over deficiency in vitamin A, iodine, zinc, vitamin B12, and many other micronutrients.

The fetus can also be overfed although that only really occurs in one condition—maternal diabetes. If the mother's glucose levels are high because of diabetes or a diabetic tendency, fetal glucose levels are also elevated and high fetal glucose will drive the fetal pancreas to release insulin. High fetal insulin levels lead to the fetus laying down fat (because insulin promotes uptake of fatty acids into fat cells) and this is why babies of diabetic mothers are usually large and fat at birth—their delivery can be problematic. These babies are more likely to be fat as children and in turn to develop diabetes, and so the cycle can be perpetuated. Because of the way in which the mother's metabolism is altered during pregnancy by placental hormones, this is a time when a pre-diabetic state is frequently exposed. There is an increasing concern about this pathway as the prevalence of obesity, pre-diabetes, and diabetes rises globally.

While we have discussed the extremes of fetal growth and nutrition, the biology of fetal development is continuous and the fetus can sense and respond to the full range of nutritional signals coming from its mother. At the extremes these can be manifest as alterations in growth, but within the normal range of signals from mother to fetus, much more subtle things are happening which can equally have lifelong consequences.

Fetal choices

Because of the way in which nutrient transfer to the fetus is regulated, it responds to the nutritional environment of its mother. The evolution of the processes underpinning the regulation of fetal development in all mammals would have been based on the probability of a good relationship between fetal nutrition and mother's nutritional environment. Selection would have focused on the normal pregnancy as pregnancies complicated by maternal ill health or placental dysfunction would in the past have had a low probability of the offspring surviving to adulthood. Thus the basic processes of development evolved so that the fetus responds to environmental signals from the mother, reflecting the state of the environment into which it will be born.

The more our research has considered the consequences of this capacity of the fetus to sense its environment, the more it became apparent that the fetus has evolved with the capacity to make some well-informed biological choices. We divide these into two types—those that give the fetus an immediate advantage and those that have an advantage much later in life. We have argued that both types of choice have adaptive value and that is why mammals have evolved with the capacity to make them both. We have termed these two types of choice short-term, or immediate adaptive responses, and predictive adaptive responses.[31]

Short-term adaptive responses are those which allow the fetus to survive an immediate environmental challenge. Some may be very transient and simply involve internal control (homeostatic) processes similar to those of adults.[32] For example if the umbilical cord kinks transiently, leading to a reduction in oxygen delivery, the fetus will conserve oxygen by reducing unnecessary movements. But many of the environmental stresses to which a fetus can be exposed exert an influence over days or weeks, and are induced for example by a sustained level of maternal nutrition or health status. Obviously if its nutrient supply is severely limited the fetus reduces its growth. It does so in an asymmetrical way by limiting blood supply to the muscles, gut, liver, and kidneys and thus protecting nutrient supply to the brain and heart. The latter is essential because if the fetal heart does not function adequately then blood cannot be pumped through the umbilical cord to the placenta, and the fetus cannot extract the nourishment and oxygen it needs.

The original observations of David Barker pointed us towards the class of developmental choice we term predictive adaptive responses.[33] We proposed that the fetus can sense its environment from maternal signals and use this information to predict—or forecast—its future postnatal environment. The availability of food is one critical environmental signal, maternal stress reflected in hormonal changes is likely to be another, but there are several other environmental signals to which the fetus or neonate can respond such as fluid deprivation, season of the year, and maternal behaviour.[34] For example in the rat, some mothers groom their babies a lot and some only a little. Those whose mothers groom them a lot grow up with different behaviours with less anxiety and a reduced stress hormone response.[35]

While extreme changes in the signals the fetus senses will induce immediate responses to ensure fetal survival, predictive responses are independent of these and can be induced by smaller changes across the full range of environments which the fetus senses. There is real advantage in trying to match the physiology we develop in our plastic phase of development to the environment we expect to inhabit. This will be more likely to give us maximal fitness and thus ensure transmission of our genes to the next generation. That is why the processes of developmental plasticity, including the capacity to utilize prediction, have evolved. Prediction is valuable in trying to improve the potential for lifelong match, because unfortunately we cannot afford to be plastic throughout our entire lives.

Many aspects of the control system for metabolism are established in early development. The settings of every part of the regulatory system—the appetite for food, the supply of food to tissues in the blood supply, the metabolic needs of different tissues such as muscle and fat, the amounts of those tissues themselves, the nervous and hormonal processes which regulate metabolism, etc.—are all influenced by developmental factors. For example in humans as in other mammals the number of muscle cells in the heart and in skeletal muscle, the number of fat storing cells in the body, and even the number of urine-forming units in the kidney (which helps control blood pressure as well as making urine) are all set before birth.

Altered gene expression induced by epigenetic changes is fundamental to these processes. Permanent changes in expression of genes involved in fat and carbohydrate metabolism and in hormone responsiveness, in part due to epigenetic modifications,[36] are found in tissues taken from the offspring of rat mothers whose nutritional state was changed during pregnancy. The complexities of these developmental cascades and their long-term consequences have barely started to be elucidated.

These processes of developmental plasticity tune further the match generated by selection. They allow us to adjust our physiological capacity to match the range of environments we anticipate confronting and thus set our comfort zone to match that predicted environment. This provides a survival advantage. In doing so it does not eradicate genetic variation, indeed it protects it[37]—here is an adaptive solution based on a phenotypic response rather than changes in the genotype.

Is the mother a reliable witness?

But the fetus can only base its predictive choices on what it forecasts the future to be and this depends on the reliability of the sensor system; how well does the information from its mother reflect the current, and particularly the future, external environment? The sensor system detects various nutrient and related signals (such as stress hormone signals) crossing the placenta. But no forecasting system can be perfect and the predictive system of fetal life is no exception. This does not matter in evolutionary terms—other things being equal, provided that it is correct more often than it is wrong, mathematical modelling shows that it will confer an advantage.[38]

There are many ways in which a fetus or neonate can be misinformed about its future. If the placenta is not working properly, for example if the mother has pre-eclampsia, irrespective of what she eats fewer nutrients will get to the fetus; that fetus will predict a poor postnatal environment and set its nutritional and metabolic physiology accordingly. The chances are increased that the physiology adopted will be mismatched with the environment that follows. This is an increasingly common scenario—the fetus predicts an adverse environment and makes decisions accordingly which leave it mismatched with the enriched environment of the child and adult.

Maternal smoking impedes nutrient supply to the fetus and leads it to predict a nutritionally deprived future. Many women, even in affluent societies, eat an unbalanced diet at the time they become pregnant and in early pregnancy. This may be because they are dieting to lose weight, because they are unaware of the composition of a balanced diet, or because of other lifestyle factors. A recent study in Southampton showed that up to 50 per cent of women were eating what was evaluated as an imprudent diet at the time of conception.[39] This was particularly the case in women of low educational attainment, highlighting the importance of education about diet, pregnancy, and fetal health. In Japan, birth weight is actually falling because women diet before pregnancy and their obstetricians believe in drastic limitations on weight gain in pregnancy because they think it reduces the incidence of pre-eclampsia.[40] In many cultures women diet in late pregnancy because they believe that this reduces the risks of difficult childbirth. Throughout the world, a high proportion of pregnancies are unplanned, so it is hardly surprising that women are often not prepared for the event. But in being so they risk misleading the fetus about the nutritional environment it will face in later life.

But even if the woman is consuming a well-balanced diet, the growth and development of her offspring is nonetheless influenced by the normal processes of maternal constraint. These operate to some degree in all pregnancies, and particularly in first pregnancies, in twin pregnancies, and in women having their first baby at the extremes (both young and old) of reproductive age. This constraint generates an upper limit on how much nutrition the fetus can sense, and thus on the range of environments it can predict. This may not have mattered prior to the development of agriculture because energy-dense environments did not exist then—but it certainly matters now.

We have suggested that maternal constraint, by dampening the nutritional information the fetus receives, may have given our species an adaptive edge during our evolution, because the predictive responses always made us expect to live in a slightly harsher environment than existed during our gestation. Thus as a species we are pre-adapted to expect worse than we may experience, and this would have given us an inbuilt safety margin, some degree of energetic reserve.[41] But as our nutritional environments have got richer, the discordance between prediction created by these constraint mechanisms and reality has got greater. Rather than being evolutionarily advantageous, these prenatal forecasts have now become disadvantageous.

A changing world

For the remainder of this chapter we focus on the very common scenario of developing in a constrained environment and then living in a richer one—one which is significantly richer in energy terms than predicted. So the individual faces a developmentally induced *metabolic mismatch*.

The mismatch can happen for several reasons: because the developing offspring has made an inappropriate prediction due to unbalanced maternal diet or illness; because maternal constraint has been followed by exposure to an affluent environment of excess nutrition and a sedentary lifestyle. Or the environment can change within the time of a single generation. The adolescent who migrates from a rural village in the Indian subcontinent to a city such as Mumbai will suddenly be potentially exposed to a nutritionally richer lifestyle associated with less physical activity. Economic migrants and refugees usually move from poorer to richer conditions. They will not be developmentally matched for the lifestyle they will then experience.

Imagine a fetus which is developing in expectation of living in a poor environment postnatally because its mother has signalled this to it. What predictive responses would it be best for that fetus to make? It will make adjustments to its development and physiology which favour successful life after birth in a nutritionally limited environment. It will reduce its muscle bulk, adjust its biology to favour laying down fat whenever it can as a form of energy reserve, and set its appetite to favour eating high-fat foods when available. It will make many other trade-offs because it will adopt a fast and furious strategy and forecast a short rather than a long life. It will form a smaller number of kidney units—predicting that it will not live long enough to need the reserve capacity. It will reduce the ability of insulin to drive glucose into muscle cells so that its demands for energy are less.[42]

In practice the changes made are very subtle; a slight change in musculoskeletal development, a small alteration in metabolic pathways in the liver,[43] and so on. In themselves such subtle changes would have little significance provided the individual lived in a environment matched to that predicted—but the reality is that increasingly we are born into a rich environment with a biology that makes it easy to lay down fat and the integrated effects of these small changes can be very significant indeed.

Most infants lose a small amount of weight and fat shortly after birth and they do not show a 'rebound' in fatness until about the age of 5. Because normal prenatal development involves a degree of maternal constraint, almost every infant experiences a degree of mismatch between the pre- and postnatal environment. The important question concerns the size and the timing of this mismatch. For this reason either excess or deficient infant nutrition can have later and surprisingly similar consequences.[44]

Through the neonatal period and infancy, nutrition is still determined by the mother and her circumstances. Milk quality is affected by maternal health. In traditional societies neonatal mortality rises where mothers are ill, and maternal death in such a society inevitably leads to infant death (sadly 500,000 women still die during pregnancy and childbirth every year worldwide). The infant remains dependent on the mother for food and support both before and after weaning. Poor infant nutrition extends the period of constraint from before birth well into infancy and such children are then primed to put on fat rapidly when food becomes available. On the other hand excess infant nutrition due to the use of cow's milk may do the same because this milk is energy rich compared to breast milk. Recent studies show that cow's milk-fed infants are much more likely to get obese as they grow older. From these observations three conclusions can be drawn. First, events in early life which involve changes in nutrition have long-term consequences because developmental plasticity and predictive responses can lead to permanent changes in metabolic control. Secondly, while we do not have adequate knowledge about neonatal nutrition infants should not be allowed either to get excessively fat or to be undernourished. Thirdly, under virtually every circumstance, breast milk is best for the baby.

Mismatched metabolism

We can see why the consequences of metabolic mismatch become manifest as heart disease and diabetes. The developmentally induced preference for high-calorie and -fat content food, coupled with the setting of the fat control system, will lead to weight gain and eventually obesity. The resistance to insulin in the presence of high levels of energy intake will begin to damage the lining of blood vessels and this allows inflammation and atherosclerotic plaques to form. This damage then links with other effects, the obesity itself, the smaller number of kidney units, and the reduced density of capillaries, which all predispose to high blood pressure and then to heart disease. The individual who forecast a poor environment but who now lives in a rich one will get obese, insulin resistant, and have high blood pressure. This group of disorders is so common that it is now known as the 'metabolic syndrome'. In people with metabolic syndrome, the risk of diabetes, heart disease, and stroke is greatly increased, and the more symptoms of the syndrome they have the shorter their life expectancy. This is increasingly the picture we see in westernized societies.[45]

The problem is particularly important among sections of the population who have inadequate education or are in lower socioeconomic groups in both the developed and developing world. Middle-class people, and those with good education, will frequently find out the best way to optimize their lifestyle—from diet to physical activity, where they live, and so on. They will devote resources that might have been used on more ephemeral things to sustaining and promoting their health and that of their offspring. They can reduce the degree of their mismatch even though their lifestyle predisposes to it. Less favoured are the poor, who like everyone else have not only inherited and developed the consequences of metabolic mismatch, but who sometimes worsen their situation through inadequate education, inability or lack of opportunity to act. The range of foods they can afford often has a higher fat and carbohydrate content. This compounds their degree of mismatch. The poor are more at risk and poverty is a major contributor to chronic disease.

What becomes manifest as disease later in life starts as subtle changes in physiology that we can detect in childhood. Increasingly the first signs of obesity appear in childhood and we can relate these to greater degrees of developmental constraint. For example being a first-born child predisposes to childhood obesity,[46] unbalanced maternal nutritional status in pregnancy is associated with occurrence of childhood and adolescent obesity[47] and to a greater risk of developing diabetes in later life.[48] But what two decades ago were subtle changes in childhood physiology are now becoming manifest as overt changes in health as the degree of mismatch rises—children are exposed to higher-energy foods and exercise less and less frequently. There is a growing pandemic of adult onset diabetes appearing in children and adolescents. A generation ago this was rare and individual cases were written up in the medical journals. Now it is common—this is the price our children pay for video games rather than soccer balls, for burgers rather than fruit and vegetables.

In many ways we are paying the price of our success as a species. The problem is not confined to developed societies, because the mismatch concept can apply across the full range of socioeconomic conditions, from poor environments switching to better in the developing world, and optimal switching to excessive in the developed world. And so we are beginning to see an epidemic of metabolic syndrome in developing societies because they are going much more rapidly through the nutritional transition that was more gradual in the developed world. In India it is estimated that the number of those suffering from hypertension will rise from an already high 118 million in 2000 to 214 million people in 2025. In the same time the number of diabetics will more than triple from 19 million to 57 million people.[49] These are enormous health burdens and have enormous personal, social, economic, and political costs.

In countries such as India, women give birth to small babies who predict a poor postnatal environment and yet face a rapid nutritional transition to a relatively westernized diet, or a richer Asian diet, one for which their developmental processes have not matched them. Some children show a different pattern but the consequences of developing diabetes later are the same. These children have mothers who themselves were born small but who have already gone through the nutritional transition and are thus more likely to develop diabetes in pregnancy. In both scenarios the starting point is poor fetal development, either of the individual or the mother, followed by a nutritional mismatch.

Why do women on the Indian subcontinent give birth to small babies? In part the answer lies in the generations of women who had poor status, leading to undernutrition and stunting in childhood. Women start child-bearing at a very young age and their nutrition before and during pregnancy is often poor; sometimes the workloads expected of them are high. Specific micronutrient deficiencies associated with vegetarianism may also contribute.[50] There are obviously fundamental social, cultural, and attitudinal issues to address if the mismatch challenge is to be met. This is such a prevalent issue that Indians were once thought to have a particular genetic risk of developing the metabolic syndrome—along the lines of Neel's thrifty genes—but it now seems more likely to reflect the particular features of this population in their particular conditions. Indeed we are now seeing the emergence of the metabolic syndrome in sub-Saharan Africans as they face the nutritional transition.

Addressing the problem

A mismatch can be addressed either by changing the prediction or by trying to correct the environment in later life to be closer to that predicted in early life. Most public health measures in the developed world focus on the latter—the promotion of healthy eating and exercise. The logic behind such an approach is compelling but it may be of limited value for many; the degree of adjustment necessary to achieve a match may be too much and therefore unrealistic and it may require pharmacological intervention rather than just lifestyle changes.

A gross misinterpretation of these ideas would be to conclude that a baby born in a particularly limited environment because of the poor socioeconomic and nutritional circumstances of its mother is successfully adapted to its environment[51] and is therefore able to cope with limited nutrition after birth. Within limits, as we and others[52] have suggested, impaired growth and reduced energy consumption are an appropriate adaptive response to a poor environment, but these responses are not without costs. If nutrition is limited, then cognitive development is impaired and the risks of infection increase. These will impact far more severely on the developing child. The distinguished nutritionist John Waterlow argued eloquently[53] that 'We should not accept a *status quo* that requires children to become stunted in order to survive and then, by labeling it as an adaptation, regard it as a respectable solution.' We agree with Waterlow—moral and ethical imperatives determine that every child in such circumstances must be given the optimal chance to grow and develop normally, but the nutritional support and energy environment must be appropriate to avoid their developing gross obesity which will generate other costs. There is an even greater imperative to find ways to intervene earlier in the life course so that fewer children are born small and at risk.

Thus the alternative, not mutually exclusive, approach is to try to promote fetal and infant health and thus change the forecast. We have argued[54] that this is likely to be a more effective and indeed a particularly appropriate point of intervention in countries such as India where there is a strong history of intergenerational stunting and of lack of societal investment or empowerment of girls and young women. The capacity for a better match following improvements in pregnancy care could be significant. There are some simple measures that would make a large difference but, while these are straightforward in principle, they are extremely difficult to apply in the context of the social and cultural milieu in which these women live. Those measures for which the evidence is reasonably compelling include delaying the age of first pregnancy until the woman's pelvis is fully grown—that is until at least four years after menarche—ensuring adequate nutritional status for young women so that they are nutritionally fit at conception and that there is adequate nutritional support and a reduction in workload during pregnancy.[55] Another apparently simple measure is encouraging people not to smoke. Smoking and the fetus are incompatible and there are good data to show that infants of smoking mothers are at greater risk of mismatch.[56]

Taking a more futuristic approach, recent experimental work[57] raises the possibility that therapeutic intervention in the newborn, when there is still some plastic capacity, might be able to change the degree of match. Newborn rats can be tricked into thinking that they are fatter than they really are by giving them injections of a hormone that is normally made by fat. This has no effect in normal pups. But in those of undernourished mothers, which were destined to get obese and insulin resistant, the hormone injection stopped the development of obesity even when the pups were fed a high-fat diet. Is this kind of strategy possible in the human?

In this chapter we have suggested that the increasing incidence of heart disease and diabetes have their origin in no small part from the mismatch that arises from the interplay between developmental plasticity and the postnatal environment. We have focused on developmental influences in determining the origin of this mismatch but there are also genetic factors. A growing number of genes have been identified which appear to alter the sensitivity of the fetus to sense or respond to environmental signals.[58] Whether any of these have been specifically selected during our evolution, as was originally suggested by Neel, is not known.

We might have hypothesized that evolutionary processes would have worked to exclude such nasty fates for our species. They have not because by and large these issues do not interfere with our reproductive fitness. These diseases until recently only appeared in middle age, well after reproduction has been completed. Evolution cannot select against traits appearing after reproduction has ceased[59] and in any event it has all happened too fast—longevity to middle age and beyond is largely a phenomenon of the twentieth and twenty-first centuries.

But while we have focused this discussion on the metabolic component of the mismatch, these changes in metabolic physiology do not occur in isolation. Setting metabolic regulation to match predicted food availability is an important component but other things happen in parallel. In a predicted poor environment the offspring can choose to invest in early reproduction and hence those born smaller tend to have earlier puberty, as we discussed in the previous chapter. The offspring will invest less in repair and maintenance because longevity is not likely to be a successful strategy and hence both animals and humans born smaller tend to have shorter lives.

Evolution has provided the tools for us to try to match our life course to the environment we predict we will face. But while this was a brilliant strategy for raising the probability of reproductive success within the environments in which we evolved, it is now failing. The degree of shift in our environment is far greater than our biology could possibly allow for without a cost and our species has been very good at changing our environment. We really have made things hard for ourselves.

8
Four Score Years and Ten

Despite the biblical references to Methuselah who is reputed to have lived until he was 969, the human who had the longest documented lifespan was the Frenchwoman Madame Calment who died in her 123rd year. Biblical or not, our best guess for the average life expectancy of our Palaeolithic ancestors is about 25 years. This is a little misleading because a large number of children died soon after birth or in infancy. Recalculated and based on the interpretation of the skeletal record it seems that, provided we survived childhood, then our average life expectancy would have been about 35–40 years.[1] This number may seem low but what is surprising is that life expectancy in relatively modern times was not so very different.

The Nobel laureate Robert Fogel, in his book *The Escape from Hunger and Premature Death*,[2] documents life expectancy in populations over the past 400 years. In England life expectancy was only 32 years in 1725 although it was 50 for those who were able to migrate to the better conditions of North America. At the time of the industrial and political revolutions at the beginning of the nineteenth century, life expectancy in England and France was still below 40. By 1900, England and France had almost caught up with their American cousins, but average life expectancy was still only 48. But by 1950 it had leapt to 68, by 1990 to 77, and it is projected to reach as much as 90 years by the year 2050. Japanese women already have a life expectancy in excess of 80 years.

Equally dramatic, although delayed, changes in life expectancy have occurred in countries undergoing rapid economic transition. In India the life expectancy was 39 in 1950, but only forty years later in 1990 it was 50. In China, life expectancy is reported to have moved from 41 to 70 over the same period. So to have large numbers within a population reaching middle and old age is a rather new phenomenon. While there have always been some individuals who reached old age, our species is really facing a dramatically new age structure.

This change in lifespan uncovers a number of potential mismatches. Were we designed to live this long—are some of the ravages of ageing a result of living beyond the evolved lifespan of our body functions and tissues? By living longer does our cumulative exposure to toxins exceed our capacity to cope with them? By living longer, women now have a much longer post-menopausal phase. Was their physiology designed to allow for this, or are there deleterious consequences of an extended period of post-reproductive life? These are all potential mismatches that arise from the reduced risk of early death which the new environments of the last 100 years have produced.

Wearing out

The last century saw the rise of the consumer culture. The new manufacturing processes as well as the new middle classes fuelled a boom in production of consumer items, from ballpoint pens to cars. A new breed of advertising professionals became adept at persuading the public to buy *their* product. In the developed world the demand seemed insatiable. Would it last? This question plagued industrialists who wanted to see continued expansion of their industries following steadily rising sales figures and continued investment. But could it last? Wouldn't there inevitably come a time when everyone had all the goods they wanted? Obviously this point must be reached eventually, and no matter how gimmicky and fashionable this year's model of car or food blender could be made to seem, there must come a point when the public desire to purchase it would become sated—unless they had to buy new toasters or cars to replace ones that had simply worn out. If these things broke down and were not repairable, then the market opportunities would be virtually unending. And so there was no advantage in manufacturers designing models to last. We all know how this works. You buy a new vacuum cleaner because it has all the features that you are sure you need—a high-powered motor and hoses which will allow unpleasant detritus to be captured in the remotest recesses of your living room, dust filters and clever mechanisms for keeping the cable wound up, and so on. It comes with a guarantee—parts and labour to be covered for twelve months, or maybe even longer. But no matter how long the period of guarantee is, the wretched thing goes horribly wrong a few weeks after that period has elapsed. No, your retailer tells you, it is no longer covered under the guarantee but . . . 'We could send it away for repair, but this may take a long time and we cannot estimate the cost. Have you considered buying a new one?'

Humans share some similarities with these consumer 'durable' items. We tend to become ill as we get older; sometimes this degeneration is not easily fixed, or costs a good deal of money; and often it seems all too predictable. The health and life insurance industry is based on those elements of predictability and likely cost. But there are two important questions. First, why do our bodies wear out? And secondly, which is really related, why is there is a period of 'guarantee of youth' during which it is unlikely that parts of our bodies will wear out? From our earlier discussions, we can see that the answers are likely to lie in thinking not only about the wear and tear of adult life, but also about developmental and evolutionary biology.

Running repairs

Development is characterized by plasticity, a period when it is possible to change the structure and function of the body not only to meet immediate demands but in prediction of future needs. But this period of plasticity has to come to an end at some point, at a time well before the individual is fully mature. It would be far too expensive in biological terms to extend plasticity indefinitely. But after this time these body parts must still be maintained, serviced, and be subject to necessary minor repairs. In the same way, the guarantee on a new car will only be honoured if we keep up the obligation to have the car serviced at the agreed mileages or times. So, to maintain a youthful body in a healthy state, it needs servicing and repairs to rectify any minor damage. The obvious examples are healing of the skin after minor cuts and abrasions and the repair of bone fractures, but the processes literally go much deeper and affect nearly every cell in the body.

These repair processes are inevitably energetically expensive, and so they cannot be kept going forever. In evolutionary terms there is little value in investment in an organism once it has completed its reproduction. For humans, our short lifespan during most of our evolutionary history meant that there would have been little if any selection pressure to support repair into old age, just as the vacuum cleaner manufacturer would not have opted to develop a model that would last indefinitely. Thus our species evolved with a life-history strategy of investing most in maintaining its younger, reproductively active members, to ensure the transmission of genetic information to the next generation.

Ultimately any car will break down—they all do. Similarly, the repair processes of the human body will ultimately decline, and we see this decline as ageing. Inevitably the decline is most evident in the tissues that have the highest maintenance requirements, those with continually dividing cells such as the skin and the lining of the gut. Repair to these becomes far less effective in the elderly. Similar processes occur in the cells lining the blood vessels, which make the vessel walls more susceptible to damage and leakage. And so we can see the inevitable failure of the body's critical systems that will lead ultimately to disease and death.

But humans are probably unique among animals in knowing the certainty of death, and we do everything within our power to delay it. We use religion to deny it through concepts of an afterlife. We no longer tolerate the evolutionary imperative of decline and demise after reproduction. We are happy that the body, like the car, has its guarantee and inbuilt repair processes operating during youth, but we also want that guarantee to last for as long as possible, long after newer replacement models are on the road. The thrust of much medicine has been to eradicate communicable disease from infections, as these kill prematurely. There have been many victories here—over smallpox, polio, and measles—but the battle continues against AIDS, influenza, malaria, and even bilharzia, as we saw earlier. In addition, there has been much attention paid to other aspects of life that promote health, such as better nutrition, clean air and water, protection from hazardous substances such as asbestos, radiation, or coal dust in the environment at home or at work. We have legislation to penalize employers who allow preventable accidents to occur to employees, and educational programmes to promote safety in the home to prevent fire or electrocution. The result has been a dramatic increase in the lifespan of humans over the past 100 years, driven originally by a progressive reduction in childhood mortality but more recently by a reduction in age-specific mortality (the chance of dying at any age) as well.

Living longer

One cost of living longer has been the rapid rise in the occurrence of diseases of degeneration and of middle and old age. These include cancer, diabetes, neurodegeneration, heart disease, and conditions which some would regard as a normal part of ageing, such as osteoporosis, osteoarthritis, and a decline in mental ability. For some of these conditions, their appearance as we age is due to the failure of repair and maintenance systems. However there is ongoing debate amongst gerontologists about the specific processes involved in ageing.[3] Either the body may simply not be able to afford additional maintenance; or the impact of progressive environmental insults on cellular function accumulates; or there is some inherent process of senescence that gives a finite life to particular tissues because they have not been designed to last. While there are multiple theories of the biology of ageing they all basically come down to this cluster of possibilities. We favour the group of theories (there are several variants on the theme) that there is a trade-off between the lifetime investment in growth, reproductive, and repair systems.[4] According to this theory those species and individuals which anticipate a short life invest less in repair and more in early reproduction and vice versa.

Once an individual has ceased reproduction and support of their progeny, there can be no selection pressures acting on them (with one possible exception we will shortly consider) and therefore there will not be selection against the inherent processes of ageing. Females have an absolute end to reproduction at menopause but reproductive capacity declines from about the age of 35, probably because the eggs which were all formed in a woman's fetal life are growing old and are less viable.

In males reproduction is possible throughout life after puberty but, while we can only speculate about Palaeolithic social structures, it is probable that male reproductive opportunity also declined with age. There are some important clues to support this idea. For example males are bigger than females and this can be taken as evidence of competitive selection for size in males. This implies that the larger and stronger males, and therefore probably younger, healthier males, had mating dominance. So again selection would not have worked to the advantage of the older male. The older male past his mating prime may have been like the old male lion, banished from the pride by the younger male, doomed to a lonely death. Alternatively, as in some primate species such as the gelada baboon, the older male remains with the colony but makes no attempt to re-enter the mating game.

In some pre-colonial societies the old and infirm were intentionally allowed to die. Yet in other societies such as the Nicobarese and some indigenous American and Australian peoples, the aged were treated with the greatest of respect.[5] In some societies old age mattered greatly in that it may have conferred the capacity of the clan to pass information through cultural inheritance with greater fidelity. Wisdom, experience, and knowledge could be important to survival under extreme conditions such as drought and this was most likely to be held in the memories of the oldest. Recent studies in the African elephant, which lives in small herds dominated by a matriarch, show that the presence of the grandmother elephant confers collective memory on the herd. It is used for example to recognize other herds of elephants as non-threatening even when previous contact with them has been infrequent.[6]

We evolved with a life-history strategy that was characterized by living in small clans at the forest edge, caring for big-brained, slowly maturing offspring. This meant that we had to space our pregnancies out appropriately. But even once her children were through infancy the Palaeolithic mother had still to support her older children until they were fully independent. A high percentage of offspring survived to reproduce—about 50 per cent, one of the highest in the animal kingdom. This strategy of having a small number of children and high parental investment led to social structures where there was some stable bonding between mother and father so that the father was also involved in child support. This was not simply altruistic because it also ensured that his genetic endowment was protected. Although the environment we lived in generally led to death from trauma, childbirth, or infection by the end of the fourth decade, in any event by then our reproductive role was effectively over.

So it is reasonable to assume that we evolved living in an environment where reproduction was largely complete by 35 years of age and life expectancy was not much longer. In such a system there would be little evolutionary pressure, and in fact potentially significant cost, to having repair systems that were effective for a period of many subsequent decades, although there would have always been some individuals who survived longer.[7] This may explain why our repair and maintenance systems become less effective in middle and old age, and it is this decline that creates one of the mismatches of ageing. We now enjoy an average a lifespan more than twice that anticipated by our Palaeolithic ancestors. The repair systems are not designed for this, and they cannot cope.

Ageing and trade-offs

Why has it worked this way? Other species can live markedly longer lives, The golden eagle can live over 80 years, the white sturgeon over 100 years, and tortoises up to 150 years. Many trees such as the giant redwoods and bristlecone pines of California and the kauri of New Zealand live for thousands of years. The creosote bush can live for more than 10,000 years.

Longevity runs in families and this suggests that genetic determinants may be involved. A study in Boston found that women who were able to conceive children naturally after the age of 40 had a four times greater chance of living to 100.[8] There are several other studies showing the relationship between a late menopause and longevity and others showing the reverse, namely that early menopause is a marker for a shorter lifespan. Such studies suggest that the capacity to reproduce late is a marker for genetic determinants of a slower tempo of life course. Experimentally in fruit flies, roundworms, and mice it is possible to select animals artificially for longevity, showing that it has indeed genetic determinants: the genes involved are those associated with growth and metabolism.[9] This finding was surprising at first but on reflection it is what we would expect. The developing organism can make trade-offs between different components of its life-course strategy in response to environmental signals and in doing so it tunes the evolutionary settings of this strategy. If it predicts a threatening environment it will invest less in growth, metabolism, repair, and longevity and try to hasten its reproduction. Conversely if it predicts a benign environment it will invest in greater longevity. Thus in mice, *prenatal* undernutrition leads to reduced longevity[10] whereas *postnatal* undernutrition leads to a marked prolongation of the lifespan.[11]

There is some indirect evidence to show that these developmental trade-offs between components of the life-course strategy also exist in humans. For example women who develop diabetes in middle age turn out to have had a significantly earlier puberty than female siblings who did not develop the disease.[12] This fits with our predictive model if the life-course strategy is chosen on the basis of forecasting a poor fetal environment but becomes mismatched in a rich environment. The fetal information led to both an earlier puberty (as detailed in Chapter 6) and a greater risk of diabetes (Chapter 7). Such a prediction would have tipped the balance towards a 'fast and furious' life course and be associated with investing less in repair and maintenance and living a shorter life. Indeed the lifespan of those who are born smaller is generally shorter.[13]

These concepts of trade-offs can also be used to understand why ageing affects particular tissues. For example, bone density is established during development as part of an adaptive strategy. Strong bones are necessary to support a large body. Bone mineral content peaks in the third and fourth decade of life and then starts to decline; this happens more rapidly in the female after the menopause. This suggests that bone strength was evolutionarily important until reproduction was completed in both sexes, but less critical at older ages. But if the prenatal environment is poor the investment in bone mineral deposition is less and the risks of developing osteoporosis in old age are correspondingly greater.[14] Thus those born small have a greater risk of osteoporosis and associated bone fracture late in life.

A tired brain

The number of brain cells laid down in fetal life is in considerable excess of the number we use as adults. There is a progressive loss of these cells from birth and throughout life because they are essentially not renewed. There are a few stem cells in the brain but the evidence that they contribute to ongoing maintenance of brain function in humans is minimal. This is in contrast to some birds where brain cells are renewed throughout life, by death of old cells and their replacement using a well-regulated process of stem cell induction.[15]

Experimentally the brains of animals which have been exposed to adverse intrauterine conditions show many alterations: there is a reduction in the number of cells in some regions, in the number of connections or synapses between them, and in the amount of nerve fibres in white matter. Recent studies using new imaging techniques in the growth-retarded human infant show that the cerebral hemispheres are smaller and the amount of grey matter is less: they appear not to catch up after birth.[16] Perhaps this is why growth-retarded infants are more likely to have later cognitive, attention, and learning deficits.[17] Does this imply that there has been a trade-off *in utero*? Does the fetus predict a dangerous and therefore shorter postnatal life and thus does not invest in a larger brain, with its greater flexibility and reserve capacity and higher metabolic demand? Whereas once such growth-retarded infants had a markedly higher chance of dying in infancy, many more now survive. Is this a mismatch which originated through a trade-off in early life but is exposed by the improvements in child survival?

This argument can be extended further—although we have to admit that it is speculative. The number of brain cells we are born with was matched by evolution to a maximum lifespan of the order of 45–50 years. But while we are living longer, we are not born with more spare brain capacity—is that why dementias appear once we exceed that age range? On the other hand there is evidence that keeping an active brain throughout life by stimulation through learning and activities such as crossword puzzles will slow the loss of brain cells—perhaps because an active brain makes growth factors which inhibit the processes of cell death.[18] This suggests that perhaps we have do have some capacity to override the cognitive impairments associated with the mismatch of ageing.

The major neurodegenerative diseases are Alzheimer's disease and Parkinson's disease. We do not know what causes them although there is some evidence that viral or toxic agents, as well as genetic factors, might be involved. But diseases associated with ageing could be induced either because of a cumulative injury throughout life or because the inherent obsolescence of the brain becomes exposed when its reserve is lost through the normal processes of ageing. These diseases are exceptionally rare in younger people and thus are a direct consequence of our living much longer, but we do not know which of the possible mechanisms are involved.

A similar idea about failure of repair can be applied to virtually every other system in the body. Ageing-related disease can be seen as the result of a trade-off between early life function and later life repair, coupled with the onslaughts of modern life taking their toll over many decades.

Longer exposure

Cancers are a problem of uncontrolled cellular growth and are also much more common as we grow older. Every time cells divide there is the risk of damage to DNA, either from mistakes in the copying process or incorrect sorting of the genetic material in the chromosomes. All cells (except red blood cells which do not have a nucleus in humans) have enzymes within them that are used to maintain the integrity of DNA and to rectify copying errors. This hidden servicing of the DNA declines with age, leaving it more prone to mistakes in copying or to damage, induced for example by oxidizing processes or toxins. And this underlies the early changes in some cancers. Inevitably they affect cells which divide most, those of the skin and those which line the gut, the lungs, the bladder, and the reproductive organs.

There is another way in which cancers may have their origin in developmental and evolutionary mismatch.[19] Some cancers appear to be caused by exposure to toxins or radiation. These are components of the modern environment to which we were not exposed when our ancestors evolved. Perhaps we do not have the necessary repair systems to deal with these new sustained exposures. The rising incidence of skin cancers including malignant melanoma in countries such as Australia can be related to more sunbathing and to the growing hole in the ozone layer of the atmosphere.

One of the functions of the liver is to detoxify substances we absorb which might otherwise poison us. Some species have evolved mechanisms to deal with specific toxins in their environment. For example the poison pea in south-west Australia has high levels of the toxin fluoroacetate. The banded hare wallaby can eat this plant happily while even the tiniest dose kills its predators, the dingo and fox and other species that would otherwise be its competitors for food.[20] The monarch butterfly feeds on milkweed which contains cardiac toxins but is not harmed by them; but this protects it from predation because it itself becomes poisonous after it eats this plant. Its predators must have learned that it forms a very dangerous meal.

We have similarly evolved toxin-clearing mechanisms based on exposures to them in the environments in which we evolved. They are not designed to allow us to cope with the host of modern chemicals to which we are now exposed. We are only beginning to understand the many developmental effects of new toxins in our environment; for example non-detoxified bisphenols derived from plastics may interfere with hormones during our early fetal development, play a role in the increasing problem of male infertility and abnormalities of the male genitalia, and contribute to the rising incidence of breast cancer.[21] Smoking is another form of exposure to novel toxins. Such a form of mismatch arises because we are exposed to chemicals that we have not evolved to detoxify.

Diet has been implicated in many cancers. Our modern diets, low in fibre and anti-oxidants, play a role in several cancer types, particularly of the colon and pancreas.[22] Our high-energy intakes also drive release of growth factors which may play a role in inducing breast and prostate cancer.[23]

Indeed larger babies are at greater risk of breast cancer.[24] One explanation is that these babies have predicted a rich environment and set their growth factor profile higher because this will give them both a survival and a reproductive advantage.

The menopause

The menopause is generally a sign of health—indicating that a woman has lived long enough to terminate naturally her reproductive phase of life. But why do women have the menopause? Studies suggest that the timing of the menopause has been relatively stable over the past century, occurring at a median age of about 50 years.[25] We can only speculate about the timing of menopause in Palaeolithic times but women of the !Kung, a hunter- gatherer group in the Kalahari desert in Africa, do have an earlier menopause at about 40.[26] Does this variation represent genetic determinants acting on timing, or does it suggest the influence of a different environment—perhaps a variant on the 'live a faster life' strategy? There is some evidence that environmental factors such as smoking or rate of early growth can influence the timing of the menopause but the size of the effect is small and, although women now smoke, the timing of the menopause has remained relatively constant in developed societies over the last 100 years.

The menopause is defined by the termination of menstruation, but generally fertility starts to fall well before then and the last cycles are not fertile. Once the ovary no longer releases viable eggs the menopause is inevitable and the cyclical pattern of hormone secretion is lost. Soon the repetitive cycle of the uterine lining growing and then being shed in response to these hormonal changes stops. The menopause thus reflects the end of a functioning ovary and the loss of oestrogen and progesterone secretion. The lack of these ovarian hormones has a number of consequences for the postmenopausal woman including a thinner skin, a dry vagina and loss of bone mineral.

The menopause is essentially unique to humans. The only other species in the wild in which a significant proportion of females have complete cessation of ovarian function is the pilot whale (*Globicephala macrorhynchus*) which appears to have menopause around 40 but a life expectancy of more than 50 years.[27] However many other species show a decline in their reproductive competence as they get older. Female African elephants show a 50 per cent decline in reproduction beyond 50 years of age, but only about one in twenty female elephants lives longer than this anyway. Rhesus monkeys also show a reproductive decline as they age beyond 20 years, but again this is roughly their lifespan.[28] Recently it has been reported that gorillas in zoos may also have the menopause.[29] But one of the difficulties is that there can be big differences in life expectancy between life in the wild and in captivity: a mouse rarely survives more than 300 days in the wild but may live over twice that long in the laboratory. Thus there may be an innate tendency to ovarian failure in animals that is not seen under natural conditions, but which is exposed if they live artificially longer when housed in zoos. Maybe we humans had much shorter lives when we evolved but now live in a zoo of our own creation.

So there are considerable difficulties of interpretation about the origins of the menopause. Is it an evolutionary accident, a form of design mismatch, which arises because we were designed for shorter lives and now live much longer? Or did evolution select the menopause because it offered us some specific adaptive advantage?

The menopause is not simply the reverse of puberty. While puberty is caused by the activation of brain-induced triggering hormones (or gonadotropins) which in turn in females induce the ovaries to function, the menopause is associated with ovarian failure while the brain triggering system is still fully functional. This leads to continued release of the gonadotropins after the menopause and these cause some of the symptoms of the postmenopausal state—for example the hot flushes. This fundamental difference between the mechanisms turning on reproductive function at the time of puberty, and turning it off at menopause might give us important clues about why the menopause occurs. Many species such as sheep, which are seasonal breeders, have no problem in turning on and off their brain triggering repeatedly every breeding season. Humans also do it twice—it is active in fetal life, inactivated in the infant and then activated again at puberty. Such considerations would suggest that if humans had evolved to have menopause, then the most efficient way to turn off reproductive function would have been to use the brain trigger mechanism once again. As this is not the case it reinforces the importance of the question: did humans evolve to have the menopause or is the menopause an incidental outcome of living longer?

Let's examine the possibility that the menopause is purely a result of living longer. The argument would run as follows. Evolution led to humans having a life-history strategy based on reproduction being completed in the third decade and this was accompanied by minimal selection pressure to live longer. Thus we evolved with a median life expectancy (excluding child mortality) of about 35. Few of us would have lived past 51 years of age, the median age of menopause in modern humans, so perhaps very few women lived long enough to experience it. As our previous discussion on ageing has suggested, there is an inherent biological trade-off between the energy expended to keep cells alive and functioning and energy devoted to reproduction. Eggs must be nurtured to stay alive by the cells surrounding them in the ovary so that they can be fit to ovulate. It seems as if we evolved so that eggs have a finite life of a maximum of fifty years, remembering that many are lost or have declined in viability before then. Sufficient survive to allow healthy reproduction until towards the end of the fourth decade and then maintenance beyond that point is energetically wasteful.

Then look at the other possibility. Certainly women show a decline in fertility from well before menopause—starting around 35 years of age. This may reflect the phenomenon of ageing eggs but there may well also be an adaptive advantage in doing so. Mothers needed to nurture their children well into childhood for them to have a higher probability of surviving to reproduce. The probability of having more surviving offspring can be shown by modelling to be enhanced if a woman stops having children in time to support the development of her youngest child. This is a better strategy in terms of the number of her surviving children than risking death with her next pregnancy and childbirth, a risk that rises with age.[30] If she continued to have children up until her death then her later children would be unlikely to survive. Thus in evolutionary terms it is better to stop having children well before the expected time of death and to invest in those already born.

An extension of this argument is the 'grandmother hypothesis'. Kristen Hawkes, an anthropologist from Utah,[31] and others have suggested that the post-menopausal period evolved in humans because there was survival advantage in grandmothers being around to support their own children in being mothers. The presence of grandmothers would make it easier for their own daughters to raise their children to adulthood. These effects may be simply to assist the mother in the practical necessities of childcare or they may involve transmission of wisdom and experience, a form of cultural inheritance. If the grandmother effect is genetically based then the grandchildren who benefited would be more likely to survive and in turn pass the genes common to them and their grandmother on to the next generation. Thus valuable characteristics could be selected through the support of women who have completed their own reproduction.

There is no doubt from studies in West Africa[32] and in French Canada[33] that the presence of a maternal grandmother does aid child survival. But the skeletal record suggests that very few people in the Palaeolithic lived beyond 45 years of age[34] and the menopause in all groups of healthy women is generally at a later age. This is evidence against the grandmother hypothesis. And mathematical modelling based on studies of a Taiwanese farming community suggests that the grandmother effect alone cannot explain the origin of the menopause—the fitness advantage conferred is not sufficient for evolution to have favoured it.[35] However any modelling on a single dataset is of limited value and cannot be seen as conclusive. When the advantage of the presence of the grandmother is combined with the need for the mother to live long enough to nurture her youngest child, and so to have no further children for some years before her death, the mathematical model does allow for the possibility that the menopause may have an adaptive origin. This is a subject of ongoing active debate.

On balance the post-menopausal period, like ageing and its consequences, has to be considered as the inevitable outcome of the mismatch between our design and how we now live. We evolved in an environment where life was short and there would have been little selection pressure to invest in systems to maintain the body for longer life. But through our ingenuity as a species we now live much longer and we find ourselves confronting a number of mismatches as a consequence. We have to find safe means to ameliorate the potential detrimental effects of these mismatches.

The post-menopausal symptoms of some women highlight the difficulties. If women did not evolve to have exposure to ovarian hormones for more than thirty-five years of their lifespan (starting from puberty), then the use of hormone replacement therapy after menopause to replace those hormones may create a further mismatch. We should not be surprised that there are health consequences with such continued hormone replacement. On the other hand if menopause is a by-product of longer life then perhaps women were designed to have oestrogenic exposure for the whole of their adult life, however long it turned out to be. We do not know the answer.

The challenges of longevity

The rapid increase in lifespan has implications for the distribution of resources in all societies. Social structures are changing as a greater percentage of the population lives well beyond retirement age and there are fewer children to later enter the workforce. Societies are struggling to find the best forms of support for the elderly—what kind of communities should they live in, what kind of activities do they need to keep them as healthy and valued contributors to society, and how can we best use the knowledge and expertise of this growing repository of human capital?

A specific problem concerns the allocation of health care resources, because greater longevity is likely to be associated with a longer period of chronic disease necessitating support, especially for those suffering from the most debilitating of degenerative diseases such as dementia. It raises the question of how such care should be paid for. Pension payments are unlikely to meet the bill. These operate by definition over a long period of time and for many in these schemes the calculations used to estimate contributions were based on assumptions of much shorter lifespan than has been the reality. There is much concern that a drastic shortfall in funds will leave many people very poorly provided for in retirement. The solution to this is not obvious. Senior citizens in many industrialized societies are already among the poorest sections of the population so the situation can only get worse. Perhaps the retirement age should be raised? This might suit some people, and it is true that the increase in longevity is associated with better health in those about to retire than even twenty years ago. Against this, we might argue that many people will not want to work for longer—they may wish to enjoy a longer retirement, and certainly most have lived their lives and made their plans in expectation of retiring at 65.

The demography of society is changing in all sorts of ways. To those of us (like the ageing authors of this book) who grew up in the youth-oriented culture of the 1960s, the change is dramatic. Politicians now have to think about how to attract the 'grey vote'. The demands within families have changed—the burden of looking after elderly family members for many more years than was the case even twenty years ago puts pressures on lifestyles and finances. There may be demands placed upon family resources by elderly relatives who, perhaps having worked all their lives to pay off a mortgage, may need to sell their property to meet the rising costs of nursing care. Yet their children may be at a stage in their lives when costs are high, in terms of university fees, etc. And the grandchildren, now young adults, may wish to place a deposit on a house of their own or raise capital to start a business. Much of the fabric of society in the West has been built on some expectations about the investment and inheritance of capital. These expectations are changing, and the resulting mismatch between expectations and reality will have to be reckoned with in some way or another.

We must end this chapter on a challenging note. There is a danger that currently ageing members of a population will divert resources from the next generation, at the risk of increasing that generation's health problems. We have seen how many of the problems of health in middle age have their origins in early life and this demands resources. But these resources are increasingly needed for the elderly. Addressing the mismatches associated with ageing is a real conundrum.

9
Match and Mismatch

The mismatch paradigm represents a shift in how we think about our place in the world. It provides a new approach to understanding some of the conditions of human existence—because in many ways we are increasingly mismatched to the world we now inhabit. The simplest conclusion we could draw is that a long, healthy life requires us to be as biologically matched as possible to the environment we inhabit, but paradoxically longevity itself brings with it further ways of being compromised. Such mismatches have been created by the very success of our species. While the history of many species has been one of evolution, population growth, then decline and extinction, our ingenuity and our unique capacity to foresee and manipulate the future should, we hope, permit us to avoid such a fate (although judged by the political response to issues such as global warming we cannot be confident).

The fundamental challenge in trying to achieve a better match is that we cannot easily override the evolutionary and developmental constraints of our constitution. The situations of the Kanuri and the Sherpa demonstrate that our innate capacities cannot overcome certain forms of naturally occurring or man-made mismatch. But we must not allow ourselves to become down-hearted about the situation. Many aspects of the problem exist precisely because human health is now better than ever before. Hygiene, public health, nutrition, and medicine allow us to live significantly longer. We have done much to control, or even to eradicate, some types of infectious disease. It seems increasingly likely that there will soon be a vaccine to protect against malaria. Many illnesses are treated by specific therapies. So we can be optimistic. After all, if humans have achieved so much, surely they can do some more? We do not have to accept that increased longevity will be associated with more years of suffering chronic disease. We should be able to get closer to everyone's ideal—to live as long as possible as healthily as possible, and then to die as peacefully as possible.

Complex processes of inheritance and development evolved to match animals and plants to their environments. We are no different, except that the evolved characteristics of our species have given us a remarkable ability to change our environments, with many consequences. A good match would imply an optimal strategy for our life course. But as we have seen, physical, nutritional, and social environments vary and change over time. The impact of our species on our environment gets greater and more far-reaching by the day. We have seen how this framework produces new ways of thinking about the human condition at several levels. New ways of thinking pose new challenges and problems, but they can also indicate novel solutions.

We have to recognize that much of the environment we inhabit is increasingly out of the optimal range for which our body's internal control systems were designed by the processes of evolution. This is exemplified in our description of metabolic mismatch. Many aspects of the environment have been dramatically altered by human action particularly in our recent past. So it is not surprising that our evolutionarily selected processes of plasticity and adaptive capacity are simply not able to cope. Our description of the mismatches associated with alterations in the timing of puberty, greater longevity, and perhaps the menopause, highlight the consequences of components of our biology operating on one, evolutionarily set, programme whilst environmental influences operate and make demands on another—it is like an orchestra attempting to play simultaneously to different conductors who insist on maintaining a different tempo.

Mismatch creates cost. The consequences of the mismatch paradigm are seen in the timing of puberty, which leads to young people being biologically mature at an age long before they are considered to be responsible adults by society. What are the consequences for their behaviour and for our attitudes to them? Our mores and attitudes to young people were moulded by a different generation in a different time, when this mismatch did not occur. The consequences of metabolic mismatch are seen in the high levels of chronic non-communicable disease such as heart disease, obesity, and diabetes that are—or are becoming—endemic in many societies. They are seen increasingly as populations include a greater number of ageing members. But far from being merely diseases of these older members, there are now high levels of obesity and related disease even in young children and teenagers. What hope do they have for future health? How will society bear the cost of their disease? The marked increase in our lifespan has exposed mismatches between our inbuilt repair mechanisms and our life course, and an extended post-menopausal period is now a common rather than uncommon event. Again, how will society cope with the increasing cost of diseases such as osteoporosis and dementia associated with this changed age distribution within the population?

The scope of mismatch

In the previous chapters we have described important examples of mismatch. There are many others. Once we begin to think in these terms we look differently at many issues affecting the human condition. To make this point without merely repeating what has been said in earlier chapters, we will briefly present three other examples. Each has important implications.

The tragedy of bottle-feeding of infants in the developing world is an example of how best intentions wrongly applied can cause mismatch. Women were persuaded by western-owned food companies that their children would do better if bottle rather than breast fed. It was promoted as a sign of social and economic advancement to bottle-feed. Well-intentioned women wanting to do the best for their children responded to this marketing. We now know that this created a mismatch which cost the health and lives of many infants. All mammals evolved to be fed until weaning by the mother and bottle-feeding with milk from another species cannot be matched to the needs of the human infant. The specific nutritional composition of mother's milk has been matched by evolutionary processes to the nutritional demands of the infant and different species have very different infant growth patterns and nutritional demands. Cow's milk is more energy and protein dense than is human milk.[1] No wonder that bottle-feeding with artificial formula or cow's milk has long-term consequences for the human infant. There is much evidence that children who have been bottle fed are more likely to get infections as infants, to develop obesity, have poorer cognitive development, and may be at greater risk of disease in later life.[2] This is a simple and preventable mismatch. Human babies should receive human milk.

Our second example illustrates how a mismatch can arise from a recent change in lifestyle or environment. The form of nearsightedness called juvenile onset or school myopia is very rare in modern hunter-gatherer societies, and yet about one-quarter of Inuit children are reported to have developed myopia when they first started going to school.[3] Why might this have happened? We know that such myopia usually appears in children at about 8 to 14 years of age, and is caused by the size of the growing eye being incorrect in relation to the power of its lens, leading to a problem in achieving a sharp focus for distant objects. Normally the growing eye responds to any change in focus on the retina by altering growth in a precise way so that focus is re-established. But in nearsighted children the control of that growth is abnormal.[4] In some populations mild myopia may now be present in over 80 per cent of adolescents. The cause is not genetic because there has been a dramatic increase in myopia over the last twenty years in Taiwanese schoolchildren—and this is genetically a very homogeneous group.[5] Instead it appears that the increased incidence of myopia is associated with urbanization and increased education—the more reading and artificial light, the more myopia.[6] So even if the predisposition to develop myopia exists in us all, and presumably will have done so in our ancestors for millennia, it is only exposed by intensive close work and/or working under artificial light as a growing child. Why is this a recent problem? Middle and distance vision were essentially all humans needed prior to the development of writing and fine machinery—our ancestors simply did not read as much as we do. But now school and homework in children 'tricks' the eye into thinking that it should set its focal length (and so its growth) primarily for close rather than distance vision. The result is that many of us now grow up mismatched, unable to clearly see much of our environment beyond this page without wearing our spectacles or contact lenses.

Thirdly, let us look at a psychological example of mismatch. Humans evolved to live in small social groups; perhaps no more than 50 to 120 people formed a Palaeolithic clan and probably most clans were smaller.[7] But now we live in enormous aggregations—sometimes tightly packed in small boxes piled on top of each other. We have social hierarchies and a mixture of interactions well beyond our evolved experience. How well are our brains matched to these challenges? To what extent might some mental illness be a reflection of a mismatch between our evolved brains and these dramatic social changes? There is a growing and respected school of psychologists who believe that such a mismatch is indeed fundamental to the ecology of mental illness.[8] However such arguments must be distinguished from the much more tenuous conclusions drawn by some workers in the discipline of sociobiology, who attempt to explain most human behaviour in neo-Darwinian terms. We would agree with those who think that the extension of adaptationist thinking into the sociobiology of human behaviour is dubious.[9] As we have discussed, there is great flexibility in behavioural development, humans live successfully in many different social structures, and our capacity for cultural learning is large.

Genes and the environment

Our thinking has moved a long way from simplistic views of the interaction between genes (nature) and the developmental environment (nurture). We view development as a series of sequential interactions between environmental factors and the phenotype at each point in development; in turn that phenotype reflects previous genetic and environmental interactions which did not just start at conception but, through the processes of epigenetic or non-genomic inheritance, extended from our parents' and grandparents' lives. Each developmental interaction may not only induce immediate phenotypic changes but also delayed responses, the effects of which depend on their predictive fidelity. So while we inherit a basic template in our genes, how we develop and function in life is much more complicated.

Some components of selection do not act on the absolute degree of a characteristic (such as ear length in rabbits) but rather on the capacity to alter that trait in response to a specific environmental stimulus (e.g. the ability to adjust ear length during development in response to the thermal environment). This is how phenotypic diversity, particularly at a physiological level, can arise from a given genotype and how even subtle changes in the environment can mould phenotypes. The real power of developmental plasticity is in the tuning of the degree of match between the environment and the organism's ability to respond to changes in that environment.

But it would be biologically inefficient for every small and transient environmental change to have long-term or irreversible consequences.[10] The sequential mechanisms of developmental plasticity provide some smoothing of environmental influences during the most plastic phase of development so that the tuning is set in response to the environment over a period of time rather than to an instantaneous cue. This smoothing effect adds to the accuracy of predictive responses, because the forecast is made on the basis of an integrated assessment of the environment. So we can envisage the genotype as providing the crude settings for the organism's development of its mature phenotype. Epigenetic and other forms of developmental plasticity drive the phenotype towards a better match. But as we have seen there are circumstances in which the resulting phenotype can turn out to be inappropriate, with deleterious consequences.

Within our generic life-course strategy, there is much individual variation. Some girls have their first period at 9 years, others at 16 years of age; some humans grow tall, others remain short; some children grow fast, some more slowly; some are lean, others become obese; some have thick bones, others thin; there are variations in the numbers of filtering units in the kidney and brain cells in the memory region of the brain; some of us have exaggerated stress responses, others have dampened responses; some will live to over 100, others will die before the age of 50. Part of this variation is genetic. Familial tall people secrete higher levels of growth-promoting hormones because of genetic differences in the control regions of the genes for these hormones and their regulators.[11] But much of the variation between individuals is induced by the processes of developmental plasticity and such variation is likely to involve coordinated changes in different life-history characteristics—we have seen the linkage between fetal development and puberty, and between fetal development and later body composition. This integrated tuning of our life-course strategy by the processes of developmental plasticity is designed to match each individual to the environment predicted, even if the adjustments are not always correct.

The mismatch paradigm

The mismatch paradigm as applied to humans is based on the same biological processes as operate in other species. This is important for two reasons. First, it means that experimental studies in animals may help us to understand our predicament, and the importance of such research needs to be publicly recognized, prioritized, and given appropriate funding. Secondly it reminds us that by changing the environment we not only create a potential mismatch for ourselves, but we also do so for other species. That humans have been responsible for the extinction of many species by over-hunting since Palaeolithic times, and that by destroying rainforests and other ecosystems we threaten the survival of many more, is well known. But the parallel has not been commonly drawn between the mismatch we have created for ourselves and that which we have imposed on other species. Even quite subtle environmental shifts can produce substantial changes in the percentage of the population of a species which falls into the mismatched category. The magnificent golden toad of the Monteverde Cloud Forest in Costa Rica was discovered in 1966 and was extinct only twenty years later. It lived underground for most of the year and surfaced during daylight only during a short period in April and May to reproduce. Why did it become extinct? The number of mistless days on the mountain had increased because of global warming. As a result when golden toads were above the ground, their very permeable skin dried out, and this was fatal.[12] Sadly many other species in Monteverde and elsewhere are affected in a similar way by the devastating human impact on the environment.

This paradigm leads to a model in which the constitution of the individual is determined both by the processes of inheritance, particularly genetic, and by developmental plasticity, including its predictive components. The consequence later in the life course is that the individual is either well matched to his or her environment or is not. The greater the degree of mismatch, the greater the risk of disease, although these need not be proportionately related. With lesser degrees of mismatch we may be able to cope at least transiently and there may be a threshold beyond which disease becomes much more likely. Many Sherpa were able to cope with iodine deficiency at the cost of some enlargement in the thyroid glands but for others, in whom a threshold of iodine deficiency had been exceeded early in development, cretinism resulted.

The end result of a potential mismatch will depend on those components of the environment to which a person is mismatched. As we saw in the previous chapters, metabolic mismatch enhances the risk of diabetes and heart disease in the face of unbalanced nutrition and a sedentary lifestyle, maturational mismatch causes problems for teenagers (or their parents depending on one's perspective) in our contemporary social environment, infant feeding mismatch increases the risk of childhood death, and so on.

An additional model of disease causation

A major lesson from this book is that human biology produces adaptive responses across the whole range of environments. The adaptive range for each individual varies within the broader range encompassing the whole species. And just as in other species adverse consequences, including disease, occur if an individual is faced with a set of environmental conditions that generate a challenge which goes beyond his or her ability to adapt fully. This perspective is radically different from the common view of the aetiology (origin) of disease, which is that either individuals are healthy until they somehow mysteriously 'develop' a disease because of the role of an external agent or that they are born with the gene for that disease. The classical causes of disease detailed in many pathology textbooks (trauma, inflammation, infection, toxic, neoplastic, drugs, metabolic, genetic, congenital, and idiopathic) give emphasis to such views of disease aetiology.

Such classical views of disease causation are illustrated most clearly by Robert Koch's concept from the later nineteenth century that major diseases such as anthrax and tuberculosis were produced by pathogenic organisms which were transmitted from an infected individual, who had the disease, to another person, who then developed the disease.[13] This model of disease was given enormous weight by the identification of organisms responsible for disease by pioneers such as Pasteur, and in a more general sense by the development of the subject of pathology—the study of the diseased body—during the late nineteenth century. The model was reinforced by the discovery of ways of preventing infectious disease by vaccination, and of treating infected individuals successfully with drugs such as penicillin, which not only saved their lives but helped to limit the further spread of the disease. When the science of epidemiology, the study of patterns of disease in populations, was developed with its emphasis on finding specific external causes, more spectacular successes in medicine were achieved. Every cigarette packet now carries a heath warning, based on the indisputable link established between smoking and cardiovascular disease, cancer and effects on fetal development.

There have been many expansive claims about how knowledge of the sequence of the human genome will change our lives. It has without doubt been one of the most fundamental knowledge revolutions in our history, but it has not yet brought solutions to specific problems—for example, the claims that we will benefit from personalized drug treatment based on an individual's genotype have yet to be demonstrated, and effective gene therapy remains elusive. However knowledge of the genome has proved of enormous value to medical science and the pharmaceutical industry and we will see the development of therapies based on application of that knowledge. But for most common diseases there is no magic bullet because there is not a single gene *causing* the disease—there is not a gene for diabetes or a gene for heart disease or a gene for Parkinson's disease.

So the idea that disease could occur in human populations as part of their *normal* biology and be induced by their interaction with their seemingly *normal* environment is relatively new. It adds a new class of causation of disease to the pathology student's list. The concept of an adaptive range for an organism is well established in comparative biology, and it is clear from studies of a wide range of species including humans that the adaptive range is very largely set by evolution and early development. From this perspective, a disease caused by a pathological degree of mismatch occurs if the individual's capacity to cope with or adapt to the environment is exceeded. This model only applies to some diseases but it does explain the pattern of those such as heart disease, diabetes, and osteoporosis where neither a single causative factor nor a single genetic cause can be found. And the genetic make-up of an individual may make the person more or less susceptible to mismatch disease. For example the outcome of a developmental mismatch leading to a greater risk of osteoporosis is dependent on the individual's particular variant of the gene coding for the vitamin D receptor.[14]

A new medical discipline

One of the central themes of this book is human evolution and development. New knowledge about the biology of development offers hope for ways to manipulate the development of the individual to achieve a better match. Furthermore an increased awareness of the ways in which environmental factors affect gene expression is essential to redress the balance in an increasingly genocentric world. Many diseases are a result of the interaction between the individual's constitution and the environment and we have seen the wide range of factors which can influence this interaction. The reality is that the genomic revolution will mean little unless we develop an understanding of how to regulate the expression of genes and explore the extent to which environmental and developmental signals can turn them on or off either permanently or transiently.

These ideas fit well with the field of ecological developmental biology or 'eco-devo', a form of biological thinking that has had a renaissance over the past decade and which is different from, although related to, the more molecular study of development known as evolutionary developmental biology or 'evo-devo'. It is the new application of eco-devo to human medicine which is at the heart of this book. Eco-devo fills the gap left by the exponents of the Modern Synthesis in evolutionary thinking in the 1930s and 1940s, who found development hard to fit into their conclusion that genetics explained the basis of Darwinian biology. The knowledge necessary to understand how genetics and development interact did not exist at that time and so they had no mechanistic framework on which to extend their ideas. Further, development by its very nature involves epigenetic changes—lifelong alterations in gene expression produced by the developmental environment and underpinned by complex biochemical processes which we are only now beginning to understand.

The complexities of development and embryology were effectively ignored by geneticists for most of the first two-thirds of the twentieth century, except in the work of Schmalhausen in Russia and Waddington in Edinburgh. They identified many of the basic principles that now underpin our understanding of how the environment influences developmental plasticity, allowing one genotype to produce multiple phenotypes and, of even more importance, how it influences the environmental range over which an individual's adaptive responses can operate successfully. Their work was largely overlooked in the enthusiasm of the post-DNA era. It is only in the last few years that biologists have again focused their attention on the interaction between evolution and development.[15]

The application of evolutionary principles in any form to medicine is relatively new. In an important book, *Why We Get Sick: The New Science of Darwinian Medicine*, published in 1994,[16] Randolph Nesse and George Williams present an adaptationist analysis of human disease. Their ideas show great insight, but remain poorly incorporated into medical thinking—many medical schools do not include evolutionary biology within their curricula. Nesse and Williams even use the term 'maladaptive mismatch' in their preface.

Our own ideas proceed a step further and take advantage of a new understanding of how the developmental synthesis adds to the modern genetic and evolutionary synthesis to produce a more complete understanding of how mismatch might develop. Perhaps the age of eco-devo medicine—and a new developmental perspective on why we get sick—is emerging.

The new science of epigenetics provides one mechanistic basis for this perspective. As we were writing this chapter, a new scientific journal, *Epigenetics*, was launched. Human epigenome projects are now under way, aimed at mapping the potential sites of epigenetic modification in our genome. An exciting new vista of research is opening up and there is much more to discover about developmental epigenetics. Why is it that some genes are more accessible to modification, how is specificity generated so that the promoter for one gene is affected while that for another is not, and what are the mechanisms by which epigenetic processes operate in specific organs during specific developmental periods? Could there be a few key genes whose regulation in early development is fundamental to inducing the particular life-course strategy that is chosen? This is an exciting area of science and it will advance rapidly in the next few years.

A life-course approach is particularly important to diseases caused by, or where the risk of developing the disease is enhanced by, mismatch. There are at least three aspects to consider: the various strands of inheritance, the environment experienced during development, and the environment now being faced. We cannot learn much about how such diseases originate, or their later consequences, simply by looking at a snapshot of the situation taken at one point in time, any more than we can tell much about where the passengers in a car have come from or where they are going by peering through the window of the stationary car. The life-course approach is of necessity a time-consuming one, and it will not be popular when governments or health administration authorities want to achieve rapid results. However the mismatch paradigm, as part of a life-course approach to understanding disease causation, can explain and even predict the patterns of health and disease which are developing rapidly, from the consequences of the dietary excesses of the western world to diseases appearing in populations in rapid economic transition. It may hold the key to interventions which could pay off over the short term and should therefore be given greater attention even by makers of short-term policy. Society will pay heavily if the life-course perspective remains marginalized.

The challenge of mismatch

So what is to be done? A logical conclusion from this book is that to improve the human condition we must increase the degree of match between the biology of members of our species and their current and future environments. This will not only make them healthier and improve their quality of life but will also reduce the risk of disease in subsequent generations.

But can we reduce the degree of mismatch we face? In some cases, such as metabolic mismatch, we clearly can; in others, such as maturational and longevity mismatches, we probably cannot but even so understanding them helps us to think of ways of minimizing their impact. Logically, to improve a match we either have to change the environment or to change the biology. The environment of the twenty-first century—even in the most deprived of modern societies—is very different from that in which pre-agricultural hominids evolved. Somewhat late in our evolutionary history the development of agriculture led to settlement. In turn this increased population density and led to new power hierarchies and social complexity. But agriculture and settlement also brought a greater risk of undernutrition and infection. These issues were magnified greatly during the industrial revolution of the eighteenth to nineteenth centuries and the technology explosion of the twentieth to twenty-first centuries. We live in an environment which in metabolic and other terms is well beyond the capacities that our biology has evolved to cope with over the last 150,000 years. One of the most important attributes we have inherited through both genetic and cultural means is our ability to change our environment. We therefore need to change our environment *again*, to achieve a better match with our biology. This is not to say that we have to revert to some neo-Stone Age existence. But it does mean that we must now focus on how the built environment of our homes and workplaces can be modified to promote the amount of exercise we take every day. We must give far greater emphasis to promoting good nutrition and access to healthier foods to allow more people to eat a balanced diet better matched to their physiology. And because these lifestyle changes are much easier for the better-off members of society, we will have to give special attention to the poor, caught in a poverty trap—because the challenges of helping them to address the problems of metabolic mismatch are so much harder.

In the book we also examined the possibility of changing the other group of factors contributing to mismatch, namely our fundamental biology. Of the components of inheritance we reviewed in Chapter 2, there does not seem much chance of changing our genes, our inherited genotype, which are the product of millennia of evolution—not until the as yet rather disappointing progress towards gene therapy advances to the point where it could be widely and ethically employed to modify the human genome. At best such approaches may address diseases such as cystic fibrosis which originate from a defect in a single gene. The possibility that they can be used to alter the multi-factorial biology which underpins the bulk of human disease seems remote.

We might however be able to modify the epigenetic component of our inheritance, as this constitutes the point of interaction between the genotype and the environment. We already know a fair amount about the processes of such epigenetic modifications, so it may be possible to develop environmental tools to alter the epigenetic expression of genes. We described one experiment in which undernourished newborn rats could be tricked into thinking they were fatter than they were: their response altered their development and made them able to withstand the potential mismatch of being fed a high-fat diet later in their lives.[17] Even relatively subtle changes in nutrition, for example adding extra folic acid to the diet in pregnancy, can have dramatic epigenetic effects, and research is ongoing to see if such processes operate in humans.

A further advantage of focusing our attention on epigenetic processes is that they can be detected early in the life course, even if their full effects do not become manifest until much later. So we can envisage measuring epigenetic markers such as methylation changes in genes in young people, or even in the placenta of a newborn baby, in order to give lifestyle advice or perhaps prophylactic therapy in time to reduce the risk of disease arising from a possible mismatch in later life.

A focus on epigenetic processes cannot be divorced from paying far greater attention to the initial components of the human life course. Some of the interventions would need to be made just before or during early pregnancy and this is a real challenge at a population level because about half of pregnancies even in developed countries are not planned. Any population-based approach would require optimizing the diet and body composition of all women of reproductive age. A further important goal should be reducing the number of teenage pregnancies—a problem of major proportions in many parts of the developing world where girls are married at puberty. We need to monitor infant and early childhood growth far more carefully, especially in children who were smaller than average or were excessively large at birth. While the biological framework for such interventions is straightforward, their application in practice to any society might be very difficult: cultural attitudes to women and children vary enormously across the globe, and in many societies they remain the most disadvantaged members of the population.

Then we must not ignore the ways in which cultural inheritance can affect future generations. What mothers, grandmothers, and aunts tell their children, grandchildren, and nieces can dramatically influence how they choose to live. Whether or not we believe that the menopause evolved to give grandmothers a particular role, there can be no doubt that mentorship plays an important part in improving or harming the quality of the maternal–infant interaction and thus influencing the child's life course. Children who are born smaller are more at risk of cognitive impairment. But if the level of maternal–infant interaction is high, the impairment can be overcome and cognitive function can be optimized.[18] Put this in the context of many societies where there is a high rate of low birth weight—in how many families is the infant swaddled and subject to minimal stimulation while the mother returns to tilling the fields or to the carpet factory?

The costs of these problems are almost unbearable, even for developed societies. Yet they come at a time when we desperately want to help developing societies to go through economic transition. We know that this transition will only increase the burden of metabolic mismatch, but we must accept that this may initially only be a secondary concern; the disadvantages of economic transition leading to increasing mismatch are outweighed by the advantages of addressing immediate issues such as malaria, HIV, food insecurity, safe water supplies, or maternal and infant mortality. But when plans are made to promote such socioeconomic transitions, we should look at the same time for ways to minimize the adverse consequences.

So economic transition can create a dilemma. In many parts of the world—the Indian subcontinent being one—some sections of the population are well advanced in the transition and are already confronting the childhood obesity, reduced exercise, and the overnutrition route to diabetes and cardiovascular disease. Simultaneously, other more disadvantaged sections of the same population suffer relative deprivation, and are born to pregnancies compromised by maternal stunting and undernutrition. Sometimes they attend the same school and are exposed to the same social intervention programmes. But whilst for the first group of children vigorous exercise and control of diet is the obvious way out, for the second it may well aggravate the problem. For girls in the latter group it could exacerbate their lack of metabolic capacity to sustain an optimal pregnancy, and so perpetuate the problem into yet another generation. The challenges of designing appropriate intervention programmes in such mixed populations will stretch the inventive capacity of any public health service. The will and, slowly, the funding for such interventions is there: but despite the urgent need to intervene, we must remember that targeted rather than global approaches will be needed.[19]

Can we evolve our way out of mismatch?

We have spent much of this book discussing the processes of evolution and how evolutionary processes have operated to develop an optimal degree of match. And there are now many examples in the biological world of evolution in action on a surprisingly fast time scale.[20] So could our own further Darwinian evolution be part of the solution to the many mismatches we face?

If there were evolutionary solutions to some forms of mismatch, they could not act to reduce those consequences of mismatch appearing in the post-reproductive period and so unfortunately they cannot provide a solution to those mismatches associated with our increased longevity. And selection can only act where the result is a change in reproductive fitness—rapid evolution would only be possible if there were a very large differential effect on survival and thus on fitness. But our skills in medicine, public health, technology, and in manipulation of the environment mean that even those mismatches which appear before or during the reproductive period will not be allowed to affect our ability to have children.

Only if the ravages of the obesity epidemic become so severe and medical science fails, with the result that young people die of the complications of the metabolic syndrome before they have families, could we envisage natural selection contributing to the way ahead—and such a scenario seems to be in the realm of science fiction. While selection would impact massively on our species if we face a catastrophic epidemic of infectious disease or a nuclear winter, the route to reducing the impact of our many mismatches lies in technological, environmental, and cultural development rather than in Darwinian evolution.

Changing our priorities

We know that there will inevitably be inertia and opposition to some of the ideas expounded in this book. Politicians are inherently sceptical of any solution that has a long time scale (a cynic would say that it has something to do with the length of the electoral cycle). The policy maker would argue that they are bombarded with requests for all sorts of interventions and the balance of priorities has to be towards the short term, where the cost–benefit ratio is easier to calculate. Indeed mothers, fetuses, and children are disadvantaged by the complexities of economic models in assessing the long-term cost–benefit of an early life intervention because such calculations depend on the so-called 'discount rate'. This is the function used by economists to assess the changing value over time of an investment made now. It can be envisaged as the opposite to the interest rate in a compound interest formula such as that used in calculating mortgage payments—in other words it represents the loss of capital growth which results from spending on health interventions now, rather than keeping the funds invested, in relation to the saving on health costs which is eventually made. The longer the time interval between intervention and benefit, the less certain we are about the level at which to set the discount rate, and we think that it has generally been set too low. Nonetheless the scientific basis for intervention is solid and the benefits, however difficult to quantify precisely, are obviously very large indeed.

A further problem is that there is a large vested interest in the pharmaceutical sector, and indeed in the electorate, in retaining the focus of health care spending on the immediate problems of diseases which become manifest in the ageing population. Because pharmaceutical companies need to recover the costs of drug development they tend to concentrate their efforts on diseases of the more affluent parts of the world where governments or individuals can afford to pay for medication. Much clinical medicine is similarly oriented, if only to keep waiting lists to acceptable levels. Even some leading medical scientists have problems with the life-course approach because it is very different from the traditional epidemiological approach. The tools of epidemiology are designed to identify the more immediate or proximate causes of disease (e.g. smoking and lung cancer) while the more distal causes (e.g. the role of smoking in pregnancy as a causative factor in osteoporosis in the offspring in later life) are difficult to identify. An exception may come from prospective cohort studies, but by definition it takes seventy years or more to quantify the link between prenatal life and for example cognitive decline in the elderly, and in such studies the relevance of what happened seventy years ago to a new generation of pregnancies is debatable. In turn this becomes a further excuse for some to ignore the conclusions already reached from research in this area.

This constitutes a strong argument for more research into the causes and consequences of mismatch. Such research will necessitate experimental studies that are robust and supported mechanistically by recent breakthroughs in epigenetic technology. It will require setting up new series of clinical studies. It may not be easy, but the implications of not redirecting investment in health research and care into this area cannot be ignored.

A fisherman speaks

Professor Zulfiqhar Bhutta from Pakistan is one of the world's most distinguished and respected public health paediatricians. He tells of visiting a poor fisherman's family in a village near Karachi. In a pleading tone, one of the fishermen said to him, 'Dr Sahib, please help us. Our children are sick, they are sick even before they are born.' If this fisherman can understand the importance of helping people to be healthy from the time before they are born, and can see it as vital and urgent, why can't health policy makers and politicians? Invoking a life-course approach and using maternal, fetal, and infant health measures to reduce the risks of chronic disease are simply not on the policy agenda—this is indefensible. The scientific logic is clear. The humanity and equity of the approach is obvious.

We will have to devote a greater proportion of our resources to promoting optimal conditions for the early years of our lives. This may involve reallocation of tax income unless new resources can be found within the GNP of developed countries. For developing countries the problem will possibly be even more acute and may involve redeploying resources from external aid agencies. In both situations it may mean reducing the relative investment deployed on our later, post-reproductive years. This will involve a major reappraisal of how human society invests its capital, and of our priorities. It will involve greater investment in younger people, who at the time may appear superficially to be healthy.

But the science we have described in this book leads to an optimistic conclusion. The mismatch paradigm does not just result from our genes: if it did it would be very hard to correct. It involves our development and the environment we have constructed for ourselves, and aspects of each of these can be changed. Many aspects of our lives *can* be improved.

Epilogue

Ang Pasang was the headman of a village situated at about 4,000 metres in a high Himalayan valley. He was a brave climber and an outstanding leader of men. He led the Sherpa team on many successful high altitude expeditions, involving climbers from all over the world. He had three sons. The first was a severe cretin who suffered cerebral palsy and gross mental retardation. Tragically he died in childhood. The second was a delightfully cheerful young man, but he was totally deaf-mute. He was also a cretin, although less severely affected than his brother. But the third son, Tsering, would be a joy for any parent—bright, intelligent, enthusiastic. What had happened? Why were two children badly affected, one more so than the other, and yet the third child was so healthy? It turned out that as Ang Pasang went on climbing expeditions he brought home foods from the expeditions—western foods with iodine in them. Gradually his wife's iodine levels rose. Her second son was less badly impaired than her first and finally, when she was pregnant with Tsering, her iodine deficiency no longer existed. So Tsering developed normally. His biological need for iodine and his access to it had been in harmony from his conception. Mismatch had changed to Match.

Notes

Introduction

1. S. B. Ortner, *Life and Death on Mt. Everest: Sherpas and Himalayan Mountaineering*. Princeton: Princeton University Press, 1999. J. F. Fisher, *Sherpas: Reflections on Change in Himalayan Nepal*. Berkeley and Los Angeles: University of California Press, 1990.
2. This form of internal constancy acting on a very short-term time scale is known as homeostasis: this concept, although not the name, was the major contribution of the great French physiologist Claude Bernard (1813–78).
3. H. K. Ibbertson, Endemic goiter and cretinism. *Clin Endocrinol Metab* 1979/8: 97–128.
4. H. B. Gibson et al., Seasonal epidemics of endemic goitre in Tasmania. *Med J Australia* 1960/47: 875–80.
5. N. E. Levine, *The Dynamics of Polyandry: Kinship, Domesticity and Population on the Tibetan Border*. Chicago: University of Chicago Press, 1988.
6. We owe this subtitle to our colleague Sir Patrick Bateson, who has written persuasively on the interplay between genes and environment in determining individual development (P. Bateson, Design for a life. In D. Magnusson (ed.), *The Lifespan Development of Individuals: Behavioral, Neurobiological, and Psychosocial Perspectives: A Synthesis*. Cambridge: Cambridge University Press, 1996) and further developed the concept in an important book with Paul Martin (*Design for a Life: How Behaviour Develops*. London: Vintage, 2000).
7. John Dupré in *Darwin's Legacy: What Evolution Means Today* (Oxford: Oxford University Press, 2003) refers to this common use of the design metaphor in evolutionary biology literature as 'Dennett's dangerous metaphor', referring to Daniel Dennett's book *Darwin's Dangerous Idea: Evolution and the Meanings of Life* (New York: Simon & Schuster, 1995). But the metaphor is well ensconced.
8. The ecologist Larry Slobodkin proposed the concepts of 'ecological time' and 'evolutionary time' (*Growth and Regulation of Animal Populations*. New York: Dover Publications, 1961). Ecological time spans just a few generations, whereas evolutionary time may be hundreds of thousand of years, adequate time for populations to evolve. Hairston and colleagues have emphasized the importance of including evolutionary processes in the analysis of how ecological systems change with time (N. G. Hairston et al., Rapid evolution and the convergence of ecological and evolutionary time. *Ecol Lett* 2005/8: 1114–27) and it is increasingly

recognized that evolution can work on a much faster time scale. For example, in the Galapagos Island finches, Peter and Rosemary Grant have demonstrated that natural selection and evolution of beak size can occur in a generation in response to climactic change caused by an El Niño event (B. R. Grant and P. R. Grant, Evolution of Darwin's finches caused by a rare climatic event. *Proc R Soc Lond B* 1993/251: 111–17). Similarly, the collapse of the cod fishery in the North Atlantic in the late 1980s and early 1990s was preceded over the previous decade by an accelerated rate of maturation of the surviving fish which was attributed to exploitation-induced evolution (E. M. Olsen et al., Maturation trends indicative of rapid evolution preceded the collapse of northern cod. *Nature* 2004/428: 932–5). In Australia, the accelerating rate of spread over the last fifty years of the cane toad, which was introduced to control insect pests but has overwhelmed native ecosystems and is toxic to predators, is associated with increased leg length in those which migrate fastest (B. L. Phillips et al., Invasion and the evolution of speed in toads. *Nature* 2006/439: 803). Evolution of tropical guppies both in their natural habitat and in the laboratory can occur rapidly (D. N. Reznick and C. K. Ghalambor, Selection in nature: experimental manipulations of natural populations. *Integr Comp Biol* 2005/45: 456–62). The rapid evolution of a new species of honeysuckle maggot arising through hybridization between two other species has been recently described (D. Schwarz et al., Host shift to an invasive plant triggers rapid animal hybrid speciation. *Nature* 2005/436: 546–9).

9. Teleology proposes that there is purpose or direction in natural processes. In evolutionary terms, implications of directionality and the use of words such as 'design', although the latter is almost inevitable, carry the danger of falling into a teleological mindset. See n. 7 above.
10. Darwin was not complimentary about the Fuegians—he described them as 'the most abject and miserable creatures I anywhere beheld' and as existing 'in a lower state of improvement than in any part of the world'. He wrote: 'These poor wretches were stunted in their growth, their hideous faces bedaubed with white paint, their skins filthy and greasy, their hair entangled, their voices discordant, and their gestures violent. Viewing such men, one can hardly make oneself believe that they are fellow creatures and inhabitants of the same world. It is a common subject of conjecture what pleasure in life some of the lower animals can enjoy; how much more reasonably the same question may be asked with respect to these barbarians. At night, five or six human beings, naked and scarcely protected from the wind and rain of this tempestuous climate, sleep on the wet ground coiled up like animals' (C. Darwin, *The Voyage of the Beagle*, 2nd edn. London: J. M. Dent, 1845). Darwin's negative view may partly have arisen from his surprise at the behaviour of the several Fuegians who were also passengers on the *Beagle*. These individuals had previously been taken to England, during which visit they were presented to the royal family, but they quickly reverted to their naked existence on their return to South America.

11. See Jared Diamond's *Collapse: How Societies Choose to Fail or Succeed* (New York: Viking, 2005); note the use in the title of 'choose' in another non-intentional context. Ecological collapse on Easter Island may have occurred even more quickly after human occupation than previously thought, according to recent radiocarbon dating (A. Gibbons, Dates revise Easter Island history. *Science* 2006/311: 1360).
12. In the 1950s, consumption of seafood contaminated with methylmercury originating from industrial waste caused an epidemic of neurological symptoms in Japanese fishermen and their families living in Minamata Bay. Toxicity was also seen in newborns as a result of transfer of methylmercury both across the placenta and into breast milk (K. Kondo, Congenital Minamata disease: warnings from Japan's experience. *J Child Neurol* 2000/15: 458–64).
13. In 1860, shortly after publication of *On the Origin of Species by Means of Natural Selection*, Darwin wrote to a friend that 'Embryology is to me by far the strongest class of facts in favour of change of forms.' The earlier embryological laws of Karl von Baer, together with his own observations on the resemblances between the larval forms of barnacles and shrimps, and between the embryos of reptiles and mammals, had helped to form Darwin's view of evolutionary divergence and of how similar (homologous) structures in organisms could arise by descent from a common ancestor.
14. Darwin's embryology (n. 13 above) has evolved into the disciplines of evolutionary developmental biology (evo-devo) and ecological developmental biology (eco-devo). Two useful introductions are B. K. Hall, *Evolutionary Developmental Biology*, 2nd edn. Dordrecht: Kluwer Academic Publishers, 1999, and S. F. Gilbert, *Developmental Biology*, 7th edn. Sunderland, Mass.: Sinauer Associates, 2003. Gilbert has also reviewed the history of how the evolutionary synthesis was slow to incorporate developmental perspectives (S. F. Gilbert, The morphogenesis of evolutionary developmental biology. *Int J Dev Biol* 2003/47: 467–77; S. F. Gilbert, Opening Darwin's black box through developmental genetics. *Nature Rev Genet* 2003/4: 736–41; S. F. Gilbert, Ecological developmental biology: developmental biology meets the real world. *Dev Biol* 2001/233: 1–12). Mary-Jane West-Eberhard's book *Developmental Plasticity and Evolution* (New York: Oxford University Press, 2003) is a comprehensive synthesis of the emergent view which gives far greater weight to developmental plasticity and phenotypic components in determining evolutionary processes.
15. This should be distinguished from a looser use of the word in developmental medicine by which one 'predicts' the future pattern of development (for example, height) from measurements taken earlier in life. In this book we use prediction to mean a form of endogenous response made in early life in expectation of a future forecasted environment.
16. P. Bateson, Fetal experience and good adult design. *Int J Epidemiol* 2001/30: 928–34.

17. In our recent book (P. D. Gluckman and M. A. Hanson, *The Fetal Matrix: Evolution, Development, and Disease*. Cambridge: Cambridge University Press, 2005) we recount the story of the Pennsylvanian meadow vole in more detail. The stimulus for the altered fur thickness is changing day length, which is detected by the fetus as changing melatonin levels in its circulation. Melatonin is secreted when ambient light is low and thus the mother's levels of melatonin are cyclical with the characteristics of the cycle changing with different day length. Melatonin is secreted for more of the 24-hour period as the days shorten towards winter and is secreted less as the days lengthen towards the summer solstice. Melatonin can cross the placenta and so the fetus also experiences the same cyclical patterns of melatonin secretion. These appear to be purely predictive responses with no immediate adaptive advantage but are clearly adaptive for the organism later in its life when it confronts the more extreme climates of summer and winter—the temperature in the nest is identical at the time the pups are born but the different coat thicknesses prepare the pup to survive the anticipated hot or cold weather once it can leave the nest (see also T. M. Lee and I. Zucker, Vole infant development is influenced perinatally by maternal photoperiodic history. *Am J Physiol* 1988/255: R831–8).
18. D. A. Krueger and S. I. Dodson, Embryological induction and predation ecology in *Daphnia pulex*. *Limnol Oceanogr* 1984/26: 219–23.
19. Darwin started *The Origin of Species* with a chapter discussing 'artificial selection', that is, selection on domestic animals for specific characteristics by farmers and bird breeders. He recognized this could change the inherited determinants of the phenotype and that similar considerations might apply in nature. He used the term 'natural selection' to describe how the impact of the natural conditions of life changed the inherited characteristics in a manner analogous to those achieved in artificial selection by the breeder.
20. Lactose is also found in human milk, but nearly all babies can produce lactase to digest this. In people with lactase deficiency, production of the enzyme ceases after early childhood and the symptoms appear at that time if milk is consumed.
21. The British surgeon James Lind observed in 1747 that citrus fruit prevented scurvy, and a regular daily ration of fresh lemon or lime juice was introduced in the Royal Navy in 1795, leading to the name 'Limeys'. But although James Cook did carry fruit juice on his ships during his Pacific explorations of 1768–79, it had been boiled to reduce its volume, destroying all the vitamin C. Cook believed, wrongly, that malt was active against scurvy, but it was probably the sauerkraut (pickled cabbage) and 'portable soup' (a vegetable extract) in his stores that actually prevented serious problems with scurvy in the crews on his long voyages. Curiously, primates are among the very few vertebrates unable to synthesize vitamin C, leading to speculation about the evolutionary significance of this mutation (I. F. F. Benzie, Evolution of dietary antioxidants. *Comp Biochem Physiol A* 2003/136: 113–26).

NOTES

Chapter 1

1. J. Weiner, *The Beak of the Finch: A Story of Evolution in our Time*. New York: Alfred A. Knopf, Inc., 1994.
2. Darwinian processes cannot act if the animal cannot reproduce, although there is at least one exception to this rule, as pointed out by the great evolutionist William Hamilton (1936–2000). In many species of insects such as ants and termites, much of the colony is comprised of infertile individuals whose role is to assist the reproductive success of their queen, to which they are closely related. Thus they ensure that their genes are passed on by assisting their kin—that is the queen—to reproduce. So here natural selection operates on the kinship rather than the individuals. There is ongoing interest amongst sociobiologists as to the importance or otherwise of kin selection in determining human behaviour. Although Hamilton did not originate the concept of kin selection, he developed 'Hamilton's rule' as a formal statement of the process. The rule relates the benefit (expressed as increased fitness of a particular gene) to the closeness of the relationship and the reproductive cost to the donor (W. D. Hamilton, The genetical evolution of social behaviour. I and II. *J Theoret Biol* 1964/7: 1–16, 17–52). A derivative of this concept is in the 'grandmother hypothesis' which focuses on the role of a grandmother in assisting the survival of her daughter's children—a matter returned to in Ch. 8.
3. P. Ball, Natural talent. *New Scientist* 29 Oct. 2005: 50–1.
4. Some species reproduce asexually, including most single-celled organisms, many plants, a few insects, and a very limited number of vertebrates. All whiptail lizards, which are desert species from the south-western USA, are female and reproduce by parthenogenesis, a process in which eggs develop without fertilization. Asexual reproduction might appear advantageous: you don't have to waste time or energy looking for a mate, and all your genes are passed on to the next generation but it reduces genetic variation. Some species, such as the water flea *Daphnia*, can reproduce asexually or sexually, and the conditions under which each strategy is used provide a clue as to the benefits of each approach. Generally, asexual reproduction occurs under conditions of low stress (for example, a stable environment with abundant food and few predators) that allows a rapid increase in population number. Conversely, sexual reproduction occurs under high stress (in our example, little food and many predators). Under these conditions, the genetic recombination that occurs during gametogenesis and the mixing of maternal and paternal alleles in the offspring lead to variation and the survival of better-adapted individuals. The geneticist William Hamilton even suggested that the primary function of the variation arising from sexual reproduction is to allow organisms to stay one step ahead of a particular form of stress—parasites. For a popular account of theories of sexual reproduction, see Matt Ridley's book *The Red Queen: Sex and the Evolution of Human Nature*. London: Viking, 1993.

5. Evolutionary biologists Stephen Jay Gould and Richard Lewontin famously used the architecture and decoration of the cathedral of San Marco in Venice, Italy, to caution against the assumption of the adaptationist school of evolution that any particular feature has evolved to perform its present function. They pointed out that while it is possible to admire how well spandrels (the approximately triangular areas between the upper parts of adjacent curved arches supporting a dome) 'function' to provide surfaces for ecclesiastical art, the very existence of the spandrels is merely a by-product of the true functional feature, the arches themselves (S. J. Gould and R. C. Lewontin, Spandrels of San Marco and the Panglossian paradigm: a critique of the adaptationist program. *Proc R Soc Lond B* 1979/205: 581–98). Indeed, 'spandrels' has been adopted as a term for features arising as by-products of evolution—such as nipples in males.
6. The mutation, named after the Australian farm where it was first observed, is in the receptor for a signalling molecule that controls the rate of development of ovarian follicles. In female sheep with the mutation, the maturation of the follicles is accelerated and more eggs are produced in each oestrus cycle (C. J. Souza et al., Bone morphogenetic proteins and folliculogenesis: lessons from the Booroola mutation. *Reprod Suppl* 2003/61: 361–70).
7. For a popular explanation, see Matt Ridley's book *Genome: The Autobiography of a Species in 23 Chapters*. London: Fourth Estate, 1999.
8. The use of the terms polymorphism and mutation differs somewhat between evolutionary geneticists and clinicians. Clinicians often use polymorphism to refer to a neutral mutation with no obvious effect and mutation where there is a clear and clinically important outcome.
9. A recent study has examined epigenetic changes in DNA methylation and histone acetylation, two 'epigenetic marks' that are known to affect gene expression during the lifetime of human twins. Differences were found that became more pronounced in older twins. Twins who had spent less of their lifetime together, or who had different health or medical histories, were more different than were twins of the same age who had spent more of their lifetime together, or who had similar health or medical histories. This study confirms that environmental exposure in development induces progressive epigenetic change (M. F. Fraga et al., Epigenetic differences arise during the lifetime of monozygotic twins. *PNAS* 2005/102: 10604–9).
10. Darwin returned from his voyage on *HMS Beagle* in 1836, and began to develop his ideas of evolution in 1837 (although he did not use the term 'evolution' until his later book, *The Descent of Man*, first published in 1871). Such ideas were then controversial from both scientific and religious perspectives, and Darwin worked slowly and in secret to perfect his theory. He was even concerned about his wife's reaction to the ideas, which impacted on her religious beliefs. In 1856, Darwin became aware that Wallace was developing similar ideas and began to prepare his own thoughts for publication. On 18 June 1858 Darwin was alarmed to receive

from Wallace, working in what is now Indonesia, a paper proposing natural selection as a mechanism for evolution. Fearing that he had been overtaken by Wallace, Darwin appealed to his scientific colleagues and friends Charles Lyell and Joseph Hooker, who proposed a joint presentation. On 1 July 1858, Wallace's paper and excerpts from an unpublished book by Darwin were read to the Linnaean Society, although Darwin could not attend because of the death of his son. Finally, Darwin published *On the Origin of Species by Means of Natural Selection* in November 1859. The first printing sold out almost instantaneously. Eventually six editions were published before Darwin's death in 1882.

11. Although Charles Darwin (1809–82) and Gregor Mendel (1822–84) were near-contemporaries, Darwin never knew about Mendel's work on inheritance in plants. Darwin published *The Origin of Species* in 1859. Although Mendel published a paper reporting his results in 1866, sixteen years before Darwin's death, it appeared in an obscure Austrian journal and its significance was not appreciated until the early twentieth century. Mendel had read a German translation of *The Origin of Species* before publishing his paper, but made no attempt to set his observations in a wider context.
12. This concept is emphasized in Kirschner and Gerhart's recent book (M. W. Kirschner and J. Gerhart, *The Plausibility of Life: Resolving Darwin's Dilemma*. New Haven: Yale University Press, 2005). They argue that selection based on transcriptional factors might be a rapid way to coordinated change in a phenotype and thus explain the evolution of novelty.
13. For an excellent series of papers summarizing current understanding about variation see B. Hallgrimsson and B. K. Hall (eds.), *Variation: A Hierarchical Examination of a Central Concept in Biology*. San Diego: Elsevier Academic Press, 2004.
14. The most widely used definition of a species invokes reproductive isolation, characterizing a species as a population that is unable to interbreed with other morphologically similar populations, but other definitions exist and there is much disagreement among biologists on definitions (not least on how to define a species in organisms that do not have sexual reproduction). In some cases one species evolves into another (anagenesis) and in other cases new species arise from a common precursor (cladogenesis). A clade is a group of species all descendent from a common precursor species. Cladistics, a concept introduced by William Hennig (1913–76), groups organisms according to shared evolved characteristics, allowing classification based on evolutionary relationships rather than morphological similarities and differences.
15. Conrad ('Hal') Waddington in Britain and Ivan Schmalhausen in Russia, working concurrently but independently in the middle of the 20th century, were the first to recognize and formulate ideas on these mechanisms. Waddington, who like so many geneticists worked with the fruit fly *Drosophila*, proposed what he called *canalization* to explain observations of the lack of variability of organisms and the

implication that minor genetic variations have minimal effects on the course of development (C. H. Waddington, *The Strategy of the Genes: A Discussion of Some Aspects of Theoretical Biology*. London: George Allen & Unwin Ltd, 1957). Waddington used the analogy of an 'epigenetic landscape', an interconnecting system of valleys, to describe how a ball (symbolizing a developing organism) rolling through this landscape might initially have several possible paths. As the valleys become deeper and narrower, development is constrained (or channelled) into one particular path (C. H. Waddington, Canalisation of development and the inheritance of acquired characters. *Nature* 1942/150: 563–5). That paper also described how disruption of canalization by more extreme conditions—a process that Waddington likened to pushing the ball over the ridge between two valleys of his epigenetic landscape—could apparently lead to the inheritance of acquired characteristics. In a related but distinct concept, Waddington observed that a particular wing structure in *Drosophila*, called *crossveinless* because the connecting veins in the wing do not form properly, could be induced by exposure of the flies to high temperature and that after several generations of selection by heat treatment *crossveinless* was expressed without exposure to high temperature. Waddington called this process 'genetic assimilation'—what was really happening was that the *crossveinless* genetic trait already existed in the flies but was selected and enriched by environmental stress until selection pressure led to it being expressed even in the absence of stress. Over fifty years later, modern biology has provided a molecular basis for Waddington's observation. The heat shock proteins or HSPs, as their name suggests, were originally discovered because they are formed in large amounts in many organisms after exposure to environmental stressors such as high temperature. HSPs are now known to act as 'molecular chaperones', binding to and protecting various critical signalling molecules in the cell; even under normal conditions, HSPs are among the most abundant proteins in cells. Rutherford and Lindquist, again using *Drosophila* as an experimental model, showed that reduction of the amount of one specific HSP, HSP90, by genetic or pharmacological means caused many different sorts of abnormalities in the flies. Selection for particular abnormalities over several generations caused them to be assimilated and expressed even if levels of HSP90 were returned to normal—an exact parallel with Waddington's observations. HSP90, presumably by its interaction with signalling molecules, appears to act as a buffer against the expression of genetic variation, and when the amount of HSP90 is reduced and buffering capacity is exceeded, the variation is expressed. More broadly, HSP90 or genes acting in a similar way can be seen as providing a mechanism for control of *evolvability*—an organism's ability to evolve—since changes in their activity will control the expression of genetic variation and in turn the opportunity for selection (S. L. Rutherford and L. Lindquist, Hsp90 as a capacitor for morphological evolution. *Nature* 1998/396: 336–42).

Schmalhausen also recognized that the normal development of organisms is protected against genetic variation and environmental change, and termed the process by which this is achieved 'stabilising selection' (I. I. Schmalhausen, *Factors of Evolution*. New York: Blakiston/McGraw-Hill, 1949). If developmental stability results from a particular optimum pattern of gene expression, then stabilizing selection will act against individuals with phenotypes that deviate from this optimum, resulting in a population with an intermediate phenotype and little variability. Stabilizing selection may be seen as the mechanism that generates the epigenetic landscape—if a particular trait has a high adaptive value, then selection for it will be stronger and the corresponding valley of canalization will be deeper; in other words, the organism is less able to tolerate variation from the norm of that trait.

Belyaev introduced the term 'destabilising selection' to describe the outcome from his fox domestication studies in which intensive selection for one trait disrupted canalization of other traits (for an accessible English-language review of this work, see L. N. Trut, Early canid domestication: the farm-fox experiment. *Amer Sci* 1999/87: 160–9). The primary effect of continuous selection for tameness appeared to be changes in stress hormone response. These hormones also control the expression of a range of other genes, particularly neural and endocrine, responsible for the timing of developmental processes. We can perhaps consider such genes that regulate the expression of other genes in development to be canalizing genes. Genetic variation in such genes may allow a coordinated change in a range of body structures and functions to be induced by a single variation.

16. The phenotypes developed may include intermediate forms as well as the two extreme phenotypes described (S. W. Applebaum and Y. Heifetz, Density dependent physiological phase in insects. *Annu Rev Entomol* 1999/44: 317–41; S. M. Rogers et al., Mechanosensory-induced behavioural gregarization in the desert locust *Schistocerca gregaria*. *J Exp Biol* 2003/206: 3991–4002).
17. The silver fox is a colour variant (about 25% of the population) of the common European and North American red fox *Vulpes vulpes*. Belyaev's study took place at the Institute of Cytology and Genetics of the Russian Academy of Sciences in Novosibirsk, Siberia, and studies are still continuing, although on a reduced scale. During this time, 45,000 silver foxes have been bred at a rate of one generation per year for a propensity to being tamed, which was carefully controlled by scheduling human contact. Now, after more than forty generations, 70–80% of animals are strongly domesticated. For further discussion and references, see n. 15 above.
18. In 1846, Charles Darwin had formulated an outline of his theory of evolution but was hesitant about its reception, in part because of his limited experience as a naturalist. Indeed, he wrote that 'no one has the right to examine the question of species who has not minutely described many'. Rebecca Stott (*Darwin and the Barnacle*. London: Faber & Faber, 2003) tells how an unusual barnacle collected

during the *Beagle* voyage prompted a study that led to eight years of work, the receipt and dissection of thousands of specimens, the publication of four books, the award of a Royal Society medal, and the acknowledgement of Darwin's status as a professional biologist.

19. R. I. M. Dunbar, *Reproductive Decisions: An Economic Analysis of Gelada Baboon Social Strategies*. Princeton: Princeton University Press, 1984.
20. These ideas draw heavily on the concepts of kin selection—see n. 2 above and for a popular account see Matt Ridley's *The Origins of Virtue* (London: Viking, 1996).
21. See n. 14 above.
22. The common ancestor of humans and chimpanzees existed about 6 to 7 million years ago and the human clade diverged from the human/chimp clade about 5 million years ago. A good candidate for an early member of the hominid clade is *Ardipithecus*. There is then strong evidence for the emergence of a number of hominid species (perhaps twenty or more), of which the gracile (slim) australopithecines such as *Australopithecus afarensis*, living in southern Africa 3 to 4 million years ago, are most likely to form part of the human lineage. The first member of our lineage to be assigned to the genus *Homo*, although somewhat controversially, was *Homo habilis* ('handy man') who has been dated to 2.3 million years ago. About 1 million years ago, *Homo habilis* was displaced by a species called *Homo erectus* (called *Homo ergaster* by some anthropologists), who were the first hominids to migrate out of Africa and whose remains have been found in southern and eastern Asia. The Neanderthals, once thought to be a subspecies of *Homo sapiens*, are now known to have diverged from modern humans at least 500,000 years ago and are placed in a separate species (*Homo neanderthalensis*)—it is most likely that they descended from Middle Eastern populations of *Homo erectus*. The other fossil hominid species found in Europe, *Homo heidelbergensis* and *Homo antecessor*, are representatives of the lineage between *erectus* and *neanderthalensis*. Between 250,000 and 100,000 years ago, *Homo erectus* in Africa was gradually replaced by *Homo sapiens*. Until recently there was much controversy about the contribution of non-African populations of *Homo erectus* to the human lineage, but molecular mapping has shown all modern humans are derived from a second African exodus of *Homo sapiens* about 65,000 years ago. Finally, we cannot help but mention the recent discovery of what appears to be the remains of a diminutive hominid, *Homo floresiensis*, on an Indonesian island, who may have evolved from *Homo erectus* and was probably made extinct by a volcanic eruption less than 20,000 years ago (P. Brown et al., A new small-bodied hominin from the Late Pleistocene of Flores, Indonesia. *Nature* 2004/431: 1055–61).
23. Recent changes in taxonomic classification mean that a 'hominid' is considered to be a member of the family Hominidae, which includes all the great apes (whose living members are humans, chimpanzees, gorillas, and orangutans), and that when specifically referring to humans and their direct and near-direct ancestors (the bipedal apes) we should use the term 'hominin'. However, the sense of

'hominid' to mean human or near-human is more likely to be familiar to readers, and we will use the term in that way while being mindful of the changed usage.

24. R. Dawkins, *The Ancestor's Tale: A Pilgrimage to the Dawn of Life*. London: Weidenfeld & Nicolson, 2004.
25. A value of 98.5% is often quoted. However, this refers to microvariation in the positions of bases which are actually present in both sequences and does not include insertion or deletion of whole blocks of sequence (R. J. Britten, Divergence between samples of chimpanzee and human DNA sequences is 5%, counting indels. *Proc Natl Acad Sci USA* 2002/99: 13633–5). Direct comparison of the human and chimpanzee chromosomes 22 found many thousand such 'indels' with major consequences for the sequences of the proteins coded (H. Watanabe, DNA sequence and comparative analysis of chimpanzee chromosome 22. *Nature* 2004/429: 382–8).
26. Roundworms, or nematodes, are perhaps the most numerous multicellular animals on earth—so numerous and ubiquitous that they led biologist Nathan A. Cobb to make his famous statement in 1915 that 'if all the matter in the universe except the nematodes were swept away, our world would still be dimly recognizable . . . represented by a film of nematodes'. Nematodes can be free-living or parasitic—there are nearly 20,000 known species and their size ranges from 0.3 mm to over 1 metre. Human parasitic nematodes have enormous medical impact, infecting an estimated 3 billion people and causing diseases such as intestinal roundworm and lymphatic filariasis (elephantiasis). Plant parasitic nematodes are economically important because of their effects on crop yields. But the nematode most familiar to biologists is the free-living soil nematode *Caenorhabditis elegans*, just 1 mm long, which rivals the mouse, the fruit fly, and the bacterium *Escherichia coli* as a 'model organism'. The attraction of *C. elegans* as a laboratory animal is its simplicity—the adult animal has just 959 cells, of which 302 cells form a simple nervous system, and how each of these cells develops has been mapped in detail.
27. The precursor to active vitamin D is in our diet and is now frequently supplemented in processed foods such as milk. But vitamin D has to be activated by the action of ultraviolet light—a reaction that occurs in the skin. Rickets is a disease of childhood caused by insufficient vitamin D in early life, and osteoporosis is the long-term consequence of inadequate bone mineral. Dark-skinned people moving to temperate climates where there is less sunlight may be at particular risk. The sunlight in the tropics was adequate to generate enough activated vitamin D to prevent rickets in their offspring, but the move to environments with less sunlight and greater body coverage with clothing puts them at a disadvantage. This is particularly the case for women wearing traditional total body coverings. Such effects may even be seen in sunny regions such as Australia (R. S. Mason and T. H. Diamond, Vitamin D deficiency and multicultural Australia. *Med J Australia* 2001/175: 236–7).

28. C. Cooper et al., Review: developmental origins of osteoporotic fracture. *Osteoporos Int* 2006/17: 337–47.
29. One of the most common ways of trying to look at genetic components of a characteristic or disease trait is to use twin studies. The concept is that if one sees a greater correlation between traits in identical twins with the same genetic makeup than in non-identical twins or singleton siblings then the magnitude of the difference implies a genetic component. This may be generally so, but identical twins also share a more common intrauterine existence than do non-identical twins because they generally share a placenta. Similarly, there are many cultural factors that lead to identical twins being treated more similarly and much has therefore been made of those studies where the twins are reared separately from birth.
30. Whether there are specific genes for plasticity or whether plasticity is implicit in the normal variability of gene expression has been the subject of lively debate. For a historical account of this discussion, see S. Sarkar, From the Reaktionsnorm to the evolution of adaptive plasticity. In T. J. DeWitt and S. M. Scheiner (eds.), *Phenotypic Plasticity: Functional and Conceptual Approaches*. New York: Oxford University Press, 2004. See also S. Via and R. Lande, Genotype-environment interaction and the evolution of phenotypic plasticity. *Evolution* 1985/39: 505–22; C. D. Schlichting and M. Pigliucci, *Phenotypic Evolution: A Reaction Norm Perspective*. Sunderland, Mass.: Sinauer Associates, 1998; M. J. West-Eberhard, *Developmental Plasticity and Evolution*. New York: Oxford University Press, 2003.
31. H. P. Guler et al., Small stature and insulin-like growth factors: prolonged treatment of mini-poodles with recombinant human insulin-like growth factor I. *Acta Endocrinologica* 1989/121: 456–64.
32. F. Aubret et al., Evolutionary biology: adaptive developmental plasticity in snakes. *Nature* 2004/431: 261–2.
33. R. S. Corruccini, An epidemiologic transition in dental occlusion in world populations. *Am J Orthodontics* 1984/86: 419–26; J. Varrela, Dimensional variation of craniofacial structures in relation to changing masticatory-functional demands. *Eur J Orthodontics* 1992/14: 31–6.
34. Queen-bee rearing occurs in three situations—when the old queen leaves with a swarm of workers to establish a new colony, when the old queen is failing and needs to be replaced, and in an 'emergency' when the queen dies suddenly or is removed (for example, by humans). In non-emergency situations, eggs that are destined to become queens are placed in special large 'queen cells' in the hive and receive royal jelly. In emergencies, some worker cells are quickly adapted and these larvae are also fed royal jelly. The workers do not raise one new queen but several, and when these emerge they must either leave the nest with some workers to form a new nest or fight the other virgin queens (and the old queen if she is present), until only one survives. The queen fights are vicious and the queens have large, early developing venom glands combined with reusable stingers to

kill their rivals. Queens may also identify rivals in their cells from olfactory and acoustic signals, and assassinate them before they even emerge. Queen fighting is unusual in social insects as most species use workers to remove extra queens, but honey bees leave it to the queens to fight it out. In a 'queenright' colony over 99.9% of workers are sterile. A few workers can lay eggs but worker reproduction is policed by other workers who eat worker-produced eggs. These are identifiable due to a lack of the pheromone that marks queen-eggs. Although it is reproductively beneficial for a worker to produce sons, a nephew (offspring of another worker) is less related than a brother (offspring of a queen) so it is in the interest of the workforce as a whole to prevent individual workers from reproducing. Thus these egg-laying workers are also attacked by other workers. In a queenless situation, up to about 50% of the workers have activated ovaries. This is due to the absence of pheromones that normally repress worker reproduction which are produced by the queen and the brood. But some investigators think that the workers refrain from their own reproduction when the colony is functioning normally, and the queen and brood pheromones are simple signals of queen fecundity.

35. See D. W. Pfennig and P. J. Murphy, How fluctuating competition and phenotypic plasticity mediate species divergence. *Evolution* 2002/56: 1217–28; S. F. Gilbert, The genome in its ecological context; philosophical perspectives on interspecies epigenesis. *Ann NY Acad Sci* 2002/981: 202–18; R. A. Newman, Adaptive plasticity in amphibian metamorphosis. *BioScience* 1992/42: 671–8; R. A. Relyea and J. R. Auld, Predator- and competitor-induced plasticity: how changes in foraging morphology affect phenotypic trade-offs. *Ecology* 2005/86: 1723–9. Denver has shown that the neurohormonal mechanism that the *Scaphiopus* toad uses to trigger accelerated development under conditions of environmental stress is very similar to that used by the mammalian fetus to initiate parturition if conditions within the uterus deteriorate, suggesting an ancient origin of this response (R. J. Denver, Environmental stress as a developmental cue: corticotropin-releasing hormone is a proximate mediator of adaptive phenotypic plasticity in amphibian metamorphosis. *Horm Behav* 1997/31: 169–79).
36. This is discussed at length in P. D. Gluckman and M. A. Hanson, *The Fetal Matrix: Evolution, Development, and Disease*. Cambridge: Cambridge University Press, 2005.
37. C. K. Williams and R. J. Moore, Phenotypic adaptation and natural selection in the wild rabbit, *Oryctolagus cuniculus*, in Australia. *J Anim Ecol* 1989/58: 495–507.
38. See Introduction n. 8.
39. S. J. Phillips and P. W. Comus, *A Natural History of the Sonoran Desert*. Berkeley and Los Angeles: University of California Press, 2000.
40. Even more problematically, an evolutionary biologist's definition of an adaptation requires proof of effects on reproductive fitness—this is a very hard test and generally cannot be achieved. Thus an adaptation is usually inferred rather than proven. In lay terminology, 'adaptation' is used more broadly to refer to

any matching response irrespective of whether it is generated by evolutionary processes. More correctly, such broader responses should be referred to as adaptedness. It may seem pedantic to the casual reader but these are important differences. When we wish to ensure that the difference is appreciated we use adaptedness and adaptation in their technical sense.

41. M. S. Croxson et al., The acute thyroidal response to iodized oil in severe endemic goitre. *J Clin Endocrinol Metab* 1976/42: 926–30.
42. A rise in temperature of 1 or 2 degrees may seem unimportant to us, but some other species are affected in ways that alter entire ecosystems. Studies in Europe and the USA have found that flowering times are steadily becoming earlier with rising temperatures. In the UK, flowering time has advanced an average of 4.5 days during the past ten years compared with the previous forty years, and some species have advanced as much as fifteen days in a decade (A. H. Fitter and R. S. R. Fitter, Rapid changes in flowering time in British plants. *Science* 2002/296: 1689–91). These changes can have significant effects on the reproductive success of some species, particularly when reproduction depends on another seasonal species such as pollinating bees. In many areas, rising temperatures are causing the migration times of insects such as butterflies to shift to earlier in the year. Birds are also breeding and migrating earlier, but in the case of species such as the pied flycatcher they have not kept up with the earlier peak in abundance of their insect prey. This mistimed reproduction has led to selection for individuals that breed even earlier (C. Both et al., Climatic effects on timing of spring migration and breeding in a long-distance migrant, the pied flycatcher *Ficedula hypoleuca*. *J Avian Biol* 2005/36: 368–73).
43. There is good evidence that control and use of fire arising from natural sources such as grass fires caused by lightning dates back at least 1 million years—the capacity to generate fire *de novo* is much more recent. Possession of fire allowed cooking of meat (making it tastier and easier to chew, which has implications for the development of human jaws and teeth) and the use of deep caves for shelter. Before then, human habitation had taken the form of rock shelters (shallow caves). The earliest evidence for built shelters dates from perhaps 500,000 years ago, with clear evidence that humans in Europe constructed quite sophisticated shelters 30,000 years ago. Although it is likely that unmodified animal pelts were used for warmth as hominids migrated from the tropics to cool temperate zones, clothing may have been a surprisingly recent innovation—a recent study of the time of divergence of the human body louse from the human head louse, an event that the investigators postulated to correlate with the introduction of clothing, indicated a date of about 70,000 years ago (R. Kittler et al., Molecular evolution of *Pediculus humanus* and the origin of clothing. *Curr Biol* 2003/13: 1414–17). In support of this, the earliest known needles, a tool specifically associated with manufacture of clothing, date from about 40,000 years ago. The earliest human tools known are crude pebble choppers dating from over 2 million years

ago, with more formed hand axes appearing perhaps 500,000 years ago and advanced flake and blade stone tools 40,000 years ago.

44. Some biologists have proposed that this process of 'niche construction' has significant implications for the evolutionary process. They argue that organisms acquire not only genes from their ancestors but also an 'ecological inheritance' consisting of the physical changes made in the environment by previous generations. Incorporating niche construction into evolutionary and ecological thinking may provide new tools for the study of complex biological systems (K. Laland and J. Odling-Smee, Life's little builders. *New Scientist* 15 Nov. 2003: 42–5; F. J. Odling-Smee et al., *Niche Construction: The Neglected Process in Evolution* (Monographs in Population Biology, vol. 37). Princeton: Princeton University Press, 2003).
45. Sir Charles Lyell (1797–1875) was a Scottish geologist whose work, particularly his 1830 book *Principles of Geology*, was largely responsible for the general acceptance of the view that the features of our present natural environment are produced by gradual changes acting through long periods of geological time, in contrast to the then dominant view of catastrophism which saw geological transformation as based on sudden change and thus allowed science to incorporate the story of the biblical flood of Noah. Charles Darwin took the first volume of Lyell's work on his voyage on the *Beagle* (1831–6) and its conclusions, as well as Lyell's approach of presenting facts and drawing broad inferences from them, strongly influenced his subsequent thought. Darwin wrote, 'The great merit of the *Principles* was that it altered the whole tone of one's mind, and therefore that, when seeing a thing never seen by Lyell, one yet saw it partially through his eyes.' Lyell did not initially accept Darwin's view of evolution, but later did so, prompting Darwin to comment, 'Considering his age, his former views, and position in society, I think his action has been heroic.'
46. The—apparent—large gaps in the fossil record are often attributed to the discontinuity of fossil-bearing sediments. But one of the implications of the punctuated equilibrium model of evolution, as proposed by Gould and Eldredge in 1972, is that these gaps are real. Gould and Eldredge suggested that instead of gradual morphological evolution the process proceeds in a series of abrupt transformations separated by long periods of stasis. Species will gradually accumulate genetic change with little change in form (which we can perhaps relate to Waddington's and Schmalhausen's ideas of canalization and stabilization). Then some, possibly external, factor will lead to a burst of morphological change and the formation of a new species. Another implication of punctuated equilibrium is that evolution occurs as the result of the formation or extinction of species, processes that may be influenced by chance events or accidents, rather than from classical natural selection. This view is clearly at odds with that of the many biologists such as Richard Dawkins who consider that evolutionary processes act in a continuous manner primarily on the gene.

47. Convention on Global Biodiversity. *Global Biodiversity Outlook*, 2nd edn., 2006. Available online from: **www.biodiv.org/doc/gbo2/cbd-gbo2.pdf**.
48. There is an excavation in Israel which suggests that contact between these two hominid species may have occurred 100,000 years ago. Presumably in Africa there was frequent coexistence of multiple hominid species and we await further discoveries from the island of Flores.
49. S. E. Moore et al., Prenatal or early postnatal events predict infectious deaths in young adulthood in rural Africa. *Int J Epidemiol* 1999/28: 1088–95; A. M. Prentice and S. E. Moore, Early programming of adult diseases in resource poor countries. *Arch Dis Child* 2005/90: 429–32.

Chapter 2

1. It is probable that RNA was the initial replicator, since in addition to being an information-coding molecule RNA can also catalyse a range of chemical reactions, including self-splicing. There may have been a primitive 'RNA world' before the appearance of DNA and protein. Later in evolution, these functions separated, with DNA becoming the replicator and proteins taking over as the catalysts. RNA still persists in our cells as the intermediate in the information flow from DNA to RNA to protein. Recent studies have revealed several previously unsuspected roles of RNA in the control of gene expression and the evolution of the genome. For discussion of the RNA world see Simon Conway Morris's book *Life's Solution: Inevitable Humans in a Lonely Universe*. Cambridge: Cambridge University Press, 2003; for a review of the role of RNA signalling in cellular regulation, see J. S. Mattick and I. V. Makunin, Small regulatory RNAs in mammals. *Hum Mol Genet* 2005/14: R121–32.
2. Jared Diamond has argued for multiple origins of agriculture with good evidence for domestication of plants and, in some regions, animals arising independently in south-west Asia, China, Central America, the Andes, and what is now the eastern USA. There is some evidence to suggest that the African Sahel, tropical West Africa, Ethiopia, and New Guinea should be added to this list (J. Diamond, *Guns, Germs and Steel*. London: Chatto & Windus, 1997).
3. Eva Jablonka from Tel Aviv has been at the forefront of those who argue that there is more to evolution than genes. A decade ago, her championing of epigenetic inheritance was largely ignored, but recent studies have provided a secure experimental foundation for the concept. See E. Jablonka and M. J. Lamb, *Epigenetic Inheritance and Evolution: The Lamarckian Dimension*. Oxford: Oxford University Press, 1995; E. Jablonka and M. J. Lamb, *Evolution in Four Dimensions: Genetic, Epigenetic, Behavioral, and Symbolic Variation in the History of Life*. Cambridge, Mass.: MIT Press, 2005.
4. See Introduction n. 8.

5. The idea of the inheritance of acquired characteristics did not originate with Lamarck: it was not even his principal idea although it is a theory now almost pejoratively known as Lamarckism. Lamarck's major idea was that the real basis of biological change was progress up a putative ladder of life with humans at the top; the inheritance of acquired characteristics was seen as a secondary process that could explain some variations. Darwin acknowledged Lamarck as an originator of the concept of the evolution of species. See Michael Ruse's book *The Evolution–Creation Struggle*. Cambridge, Mass.: Harvard University Press, 2005.
6. See n. 3 above.
7. R. E. Simmons and L. Scheepers, Winning by a neck: sexual selection in the evolution of giraffe. *Am Nat* 1996/148: 771–86.
8. It is important to distinguish sexual selection, mediated by reproductive competition between individuals of the same sex, from sexual conflict, which arises from the different reproductive investments of males and females. See n. 24 below and Ch. 3 n. 4, and G. Arnqvist and L. Rowe, *Sexual Conflict*. Princeton: Princeton University Press, 2005.
9. The issue of sexual selection was one of the intellectual differences between the two discoverers of evolution by the process of selection. Darwin spent much time in *The Descent of Man* discussing whether animals had the equivalent of an aesthetic sense and believed that exaggerated and costly male traits evolved because the female based her mating choice on aesthetics. But Wallace, who understood the argument well, came to believe that most male-specific features such as a larger body size or antlers or the lion's mane were selected because of their value in competition between males. Current opinion sees both an aesthetic and a utilitarian component. The peahen is responding to those characteristics of the peacock's tail such as size and symmetry which are good indicators of strength and health and therefore of fitness. From the female's point of view, this is both utilitarian (the outcome is healthier chicks) and aesthetic (she has been attracted by the features of the tail).
10. The Irish elk played a small part in the history of evolutionary theory, by being one of the species no longer living but found only as fossils which allowed the French zoologist Georges Cuvier to establish in 1812 that extinction is a fact. The idea of extinction had previously been strongly contested by those who believed that a creator of a perfect world would not allow whole species to disappear. The story is told in full elsewhere (S. J. Gould, The misnamed, mistreated, and misunderstood Irish elk. In *Ever Since Darwin*. New York: W. W. Norton, 1997).
11. M. Potts and R. Short, *Ever Since Adam and Eve: The Evolution of Human Sexuality*. Cambridge: Cambridge University Press, 1999.
12. The alternative views are that the loss of body hair was an adaptation to assist heat loss, or that less hair may help protect against skin disease and parasites. See Jonathon Kingdon's book *Lowly Origins: Where, When and Why our Ancestors First Stood Up*. Princeton: Princeton University Press, 2003.

13. The Palaeolithic (Old Stone Age) period is generally considered to begin around 2 million years ago with the introduction of stone tools by hominids such as *Homo habilis* and to end with the introduction of farming around 11,000 years ago at the beginning of the Neolithic (New Stone Age) period. The Neolithic ends with the widespread adoption of metal tools 4,000 to 3,000 years ago (Bronze and then Iron Ages).
14. The defect is in the enzyme 5α-reductase that converts testosterone to dihydrotestosterone. But their testis can make the other hormone, Mullerian inhibitory factor, which stops the internal female features (uterus and cervix) developing. However dihydrotestosterone is needed to make the external genitalia into the male form and ordinary testosterone cannot do so (without dihydrotestosterone the penis does not develop properly and the scrotum is not properly formed); the male children are born looking more like a girl, but with a big clitoris (the underdeveloped penis) and thick labia (the two halves of an unfused scrotum). But internally they do not have a uterus or cervix. At puberty the testis starts making much more testosterone than it could as a fetus and even in its weaker form the amount made is sufficient to make the penis grow and to induce development of pubic and facial hair (I. Imperato-McGinley, Androgens and male physiology: the syndrome of 5α-reductase-2 deficiency. *Mol Cell Endocrinol* 2002/198: 51–9).
15. See Introduction n. 19.
16. C. Darwin, *The Variation of Animals and Plants under Domestication*. London: John Murray, 1868.
17. Darwin struggled with this gap in his understanding and attempted to come up with a theory of inheritance, what he termed panglossia, an unsatisfactory part of his writings. It contrasts with the insights he developed when he based his ideas on his firm knowledge of what had been observed in the natural world.
18. The acceptance of the Modern Synthesis was preceded by a vigorous debate in the early years of the 20th century between two schools of thought about the nature of variation. The Mendelians, led by William Bateson, argued for variation occurring in the discontinuous manner typified by Gregor Mendel's peas—green or yellow, round or wrinkled—whereas the biometric school, represented by Karl Pearson and Walter Weldon, pointed to much other experimental evidence indicating continuous variation. The beginning of the reconciliation between the two schools was the demonstration by Ronald Fisher in 1918 that Mendelian inheritance acting across *many* genes could account for continuous variation.
19. Eugenics refers to studies of the genetic improvement of human populations by use of social policies. The term originates with the English statistician Francis Galton, a cousin of Charles Darwin, in the late 19th century. Eugenic programmes were instigated in many countries in the early 20th century, but the scientific foundations of these programmes (a biological basis for race and the belief that single genes are responsible for the majority of human physical, mental, and behavioural traits) are now accepted to be false. Eugenics has never

recovered from its association with the death camps of Nazi Germany, and even today any biomedical research that offers the possibility of modifying human genes is considered highly controversial. For further reading, see Diane B. Paul's collected essays on eugenics (*The Politics of Heredity: Essays on Eugenics, Biomedicine, and the Nature–Nurture Debate*. Albany, NY: SUNY Press, 1998).

20. Two of the methods by which gene expression may be modified by epigenetic changes are DNA methylation and histone acetylation. The first stage of gene expression involves binding of various transcription factors to DNA sequences, called promoters, adjacent to the actual gene coding regions. Promoters are commonly rich in cytosine-guanosine (CpG) dinucleotides, and the process of DNA methylation involves transfer of methyl groups to these cytosine residues, a reaction catalysed by an enzyme called DNA methyltransferase. Methylation of the CpG islands in the promoter will 'silence' gene expression by interfering in some way with its ability to bind transcription factors. Histones are small proteins that package the DNA in cells into bead-like structures called nucleosomes, formed by regular wrapping of the DNA around the histone core. Chemical modification of these histones can change the tightness of the wrapping of DNA around the core and increase or decrease the accessibility of the DNA to the various factors involved in gene expression. For example, acetylation of histones stimulates gene expression by relaxing chromatin structure, whereas deacetylation has the opposite effect. Epigenetic marking of DNA in these ways appears to have roles in imprinting, a non-genetic method by which information is transferred between generations, and in the development of cancer. The chemical processes of epigenetic marking are dependent on adequate supplies of the required substrates and cofactors, and effects of diet on imprinting and cancer development are well established. For a discussion of the role of methylation in genomic imprinting, see J. F. Wilkins, Genomic imprinting and methylation: epigenetic canalization and conflict. *Trends Genet* 2005/21: 356–65 and for an account of epigenetic changes as a target in cancer therapy see J. Gilbert et al., The clinical application of targeting cancer through histone acetylation and hypomethylation. *Clin Cancer Res* 2004/10: 4589–96. For a discussion of the different ways of changing DNA methylation during development and of the role of another epigenetic marking system, Pc-G/trx protein complexes, that mediates epigenetic effects in *Drosophila* and possibly also mammals, see A. Bird, DNA methylation patterns and epigenetic memory. *Genes Dev* 2002/16: 6–21.
21. R. Ohlsson et al., IGF2 is parentally imprinted during human embryogenesis and in the Beckwith–Wiedemann syndrome. *Nature Genet* 1993/4: 94–7.
22. P. N. Schofield et al., Genomic imprinting and cancer; new paradigms in the genetics of neoplasia. *Toxicol Lett* 2001/120: 151–60.
23. Prader–Willi syndrome and Angelman syndrome are human genetic diseases that affect physical and/or neurological development and are caused by loss of function of imprinted genes on chromosome 15. Some cases of these disorders

have been shown to be caused by imprinting defects involving the same gene locus—Prader–Willi syndrome involves loss of paternal gene expression because of the aberrant presence of a maternal imprint, whereas Angelman syndrome involves loss of maternal gene expression because of the aberrant presence of a paternal imprint (K. Buiting et al., Epimutations in Prader–Willi and Angelman syndromes: a molecular study of 136 patients with an imprinting defect. *Am J Hum Genet* 2003/72: 571–7).

24. The genetic conflict theory proposed by David Haig and colleagues suggested that in mammals the interests of males and females in their offspring may not coincide—the father's interest is to maximize the size of the offspring by extracting as many resources as possible from the mother, whereas the mother's interest is to conserve resources for subsequent pregnancies and for her own survival. Haig suggested that this conflict would be mediated by imprinting of genes in the fetus—and indeed, there seems to be a tendency for paternally expressed imprinted genes to promote fetal growth whereas maternally expressed imprinted genes restrain fetal growth. An example in some mammals is the growth factor IGF-2, which is produced by the fetus from the paternal copy of its gene. A molecule called IGF-2 receptor, which 'mops up' and inactivates IGF-2 and so tends to restrain growth, is produced from the maternal copy of its gene (D. Haig and C. Graham, Genomic imprinting and the strange case of the insulin-like growth factor II receptor. *Cell* 1991/64: 1045–6; D. Haig, Genetic conflicts in human pregnancy. *Q Rev Biol* 1993/68: 495–532). This conflict theory is by no means generally accepted and the imprinting of the IGF-2 receptor is not universal—for example, in humans the IGF-2 gene is imprinted but that for the receptor is not.
25. The most obvious form of gene silencing occurs in the sex chromosomes. In mammals, females have two X chromosomes (one inherited from each parent) and males only one—instead males have a Y chromosome inherited from their father. The Y chromosome is much smaller than the X chromosome and there are only a few genes that the two forms of sex chromosome have in common. For most of the genes on an X chromosome only one copy is sufficient and so after some initial cell divisions the female embryo silences one copy of the entire X chromosome in every cell. This silencing also involves chemical restructuring of the DNA including DNA methylation, although there is debate about whether this targets sections of the genome which are already transcriptionally silent because of other processes.
26. V. L. Roth, Inferences from allometry and fossils: dwarfing of elephants on islands. *Oxf Surveys Evol Biol* 1992/8: 259–88.
27. Vernalization is the promotion of flowering in some plants by exposure to a period of cold weather. Recent data suggest that a plant's 'memory' of exposure to winter is mediated by epigenetic gene silencing (R. Bastow et al., Vernalization requires epigenetic silencing of FLC by histone methylation. *Nature* 2004/427: 164–7).

28. Epigenetics must be distinguished from 'epigenesis', which is an obsolete term referring to the concept that an organism develops from an undifferentiated form by progressive development—as opposed to the preformationist view that each individual is created in its final form and only grows during development. Until about 200 years ago the preformationist belief that the sperm was a miniature version of the infant and that the egg merely provided the nourishment was widely held, including by the discoverers of spermatozoa, the Dutch microscopists Antoni van Leeuwenhoek (1632–1723) and Nicolaas Hartsoeker (1656–1725). Eventually this preformationist concept of the homunculus, as symbolized by Hartsoeker's 1694 drawing of a tiny man in a human spermatozoon, was eradicated by better microscopy and the science of embryology. For an account of the history of this debate, including the controversy among the preformationists of whether the sperm or the egg held the preformed individual, see Carla Pinto-Correia's book *The Ovary of Eve: Egg and Sperm and Preformation* (Chicago: University of Chicago Press, 1997).
29. One way in which environmental factors such as toxins can induce cancer is by epigenetic change in tumour suppressor genes that normally prevent unrestrained growth in cells (J. Gilbert et al., The clinical application of targeting cancer through histone acetylation and hypomethylation. *Clin Cancer Res* 2004/10: 4589–96).
30. P. Cubas et al., An epigenetic mutation responsible for natural variation in floral symmetry. *Nature* 1999/401: 157–61.
31. R. A. Waterland and R. L. Jirtle, Transposable elements: targets for early nutritional effects on epigenetic gene regulation. *Mol Cell Biol* 2003/23: 5293–300.
32. M. D. Anway et al., Epigenetic transgenerational actions of endocrine disruptors and male fertility. *Science* 2005/308: 1466–69. See also Ch. 7 n. 30.
33. A. J. Drake and B. R. Walker, The intergenerational effects of fetal programming: non-genomic mechanisms for the inheritance of low birth weight and cardiovascular risk. *J Endocrinol* 2004/180: 1–16; A. J. Drake et al., Intergenerational consequences of fetal programming by in utero exposure to glucocorticoids in rats. *Am J Physiol* 2005/288: R34–R38.
34. L. H. Lumey, Decreased birthweights in infants after maternal *in utero* exposure to the Dutch famine of 1944–1945. *Paediatr Perinat Epidemiol* 1992/6: 240–53.
35. L. Ibanez et al., Reduced uterine and ovarian size in adolescent girls born small for gestational age. *Pediatr Res* 2000/47: 575–7.
36. Some critics would disagree with the use of the term 'inheritance' in this context, on the grounds that for humans much of what is passed between generations culturally involves a degree of conscious choice and selection by each successive generation. Using the term inheritance might be misleading if it downplayed the importance of the enormous potential for acquired behaviour and choice throughout human life. For definitions and discussion of types of inheritance see E. Jablonka and M. J. Lamb, *Evolution in Four Dimensions: Genetic, Epigenetic,*

Behavioral, and Symbolic Variation in the History of Life. Cambridge, Mass.: MIT Press, 2005.

37. M. Kawai, Newly-acquired pre-cultural behavior of the natural troop of Japanese monkeys on Koshima Islet. *Primates* 1965/6: 1–30; K. Watanabe, Fish: a new addition to the diet of Japanese macaques on Koshima Island. *Folia Primatologica* 1989/52: 124–31.
38. R. Sapolsky, *A Primate's Memoir: Love, Death and Baboons in East Africa*. London: Jonathan Cape, 2001.
39. J. de Boinod, *The Meaning of Tingo: And Other Extraordinary Words from Around the World*. London: Penguin 2005.
40. See M. Erard, A language is born. *New Scientist* 22 Oct. 2005: 46–9, and for a more formal description see W. Sander et al., The emergence of grammar: systematic structure in a new language. *Proc Natl Acad Sci USA* 2005/102: 2661–5.
41. The emergence of language provided the ancestors of modern humans with an adaptive advantage over other coexisting hominid species. But when, and how, did language appear? And in discussing these issues, should we consider language and speech separately? After all, many hearing impaired humans communicate successfully by sign language. The palaeontological record is not helpful in this context. There is no sudden and dramatic increase in hominid brain size that can be correlated with the emergence of language, so the change appears to be associated with brain wiring rather than capacity. Similarly, although speech requires certain anatomical structures such as vocal cords, larynx, and tongue, these soft tissues are not preserved in fossils. However, speech-associated structures can leave imprints on the skeleton, and there are some indications from fossil skeletons that *Homo erectus* of about 1 million years ago might have had primitive speech. But modern speculations about the origins of language take us further back in time to the development of bipedalism about 4 million years ago—the hominid hand was now available to make gestures. We can perhaps envisage that successful communication by gestures and facial expressions—'Lion! Over there!'—selected for brain structures facilitating such actions and allowed their use for symbolic representation—at a later date, vocalizations, and even later speech, would have been added to the repertoire of communication methods. Supporters of this theory point to the existence of a brain function in primates called the 'mirror system' that is involved in both making and seeing gestures. This function can be localized to a specific part of the brain, and in humans the homologous structure is called Broca's area, which is also known to be involved in speech production—providing a link between gesturing and speech (M. A. Arbib, From monkey-like action recognition to human language: an evolutionary framework for neurolinguistics. *Behav Brain Sci* 2005/28: 105–24). Archaeological evidence suggests that the evolution of language and speech was essentially complete by perhaps 40,000 to 30,000 years ago, because at that time there was a 'cultural explosion' in, for example, tool-making, art, and social practices that can

be best explained by efficient communication among the humans living at that time. Finally molecular geneticist colleagues are beginning to get involved in this area of neurolinguistics. A family with an inherited defect in speech has been shown to have a mutation in a gene for a transcription factor called *FOXP2*, and a brain scan study has linked this defect to underactivity in Broca's area. And, intriguingly, *FOXP2* in humans has evolved rapidly since the hominid clade split from the chimpanzee lineage (M. C. Corballis, *FOXP2* and the mirror system. *Trends Cogn Sci* 2004/8: 95–6).

42. I. Tattersall, *Becoming Human: Evolution and Human Uniqueness*. New York: Harcourt Brace & Co., 1998.
43. D. S. Wilson, *Darwin's Cathedral: Evolution, Religion and the Nature of Society*. Chicago: University of Chicago Press, 2002.
44. See n. 2 above.
45. R. Sear et al., Maternal grandmothers improve nutritional status and survival of children in rural Gambia. *Proc R Soc Lond B* 2000/267: 1641–7.
46. E. A. Mitchell et al., Reduction in mortality from sudden infant death syndrome in New Zealand: 1986–92. *Arch Dis Child* 1994/70: 291–4.

Chapter 3

1. The process of retaining juvenile characteristics in adults is called paedomorphosis. In general, newts and salamanders live in aquatic environments as the juvenile gilled form and then metamorphose into adult forms that can live in both aquatic and terrestrial environments. However, the life-history strategies of particular species can vary widely from this typical model. Some species exist only as gilled, but mature, aquatic forms (obligate paedomorphosis) whereas others develop directly into the adult form without a gilled stage. Others follow the canonical strategy of a juvenile form always followed by an adult form (obligate metamorphosis). Along with several other species, the alpine newts have the ability to 'choose' their life history according to environmental conditions, resulting in a mixture of mature aquatic adults and metamorphosed terrestrial adults in the same population (facultative paedomorphosis) Facultative paedomorphosis can be interpreted as risk spreading, as the two forms can live in different niches in their environment and the species is therefore more able to resist changes in factors such as food supply or presence of predators (M. Denoel et al., Evolutionary ecology of facultative paedomorphosis in newts and salamanders. *Biol Rev* 2005/80: 663–71).
2. R. Dawkins, *The Ancestor's Tale: A Pilgrimage to the Dawn of Life*. London: Weidenfeld & Nicolson, 2004.
3. In this situation of polyandry, the best strategy for the male is, in the words of Confederate General Nathan Bedford Forrest, to 'get there firstest with the

mostest'. That done, his problem is to prevent sperm from subsequent copulations with rivals from fertilizing the female, and one tactic to do that is semen coagulation, where the ejaculate forms a plug that prevents other sperm from passing up the female's reproductive tract. In primates, sperm coagulation involves a protein called semenogelin that is produced by the prostate gland. Semenogelin has evolved much faster in chimpanzees, where each female has several male partners in each ovulatory period, than in gorillas, where each female has only one partner, suggesting that semen coagulation is important in sexual competition in primates (S. Dorus et al., Rate of molecular evolution of the seminal protein gene *SEMG2* correlates with levels of female promiscuity. *Nature Genet* 2004/36: 1326–9). For other examples of sexual competition, see Tim Birkhead's book *Promiscuity*. London: Faber and Faber, 2000.

4. More examples of human sexual strategies and behaviour can be found in Robin Baker's book *Sperm Wars: Infidelity, Sexual Conflict and Other Bedroom Battles*. London: Fourth Estate, 1996, and in Jared Diamond's *Why Is Sex Fun? The Evolution of Human Sexuality*. New York: Basic Books, 1997.
5. François Jacob, André Lwoff, and Jacques Monod won a Nobel Prize in 1965 for their discovery of operons, sets of related metabolic genes together with their molecular switches, in bacteria. We now know that genes in multicellular organisms are packaged and regulated in very different and much more complicated ways, but at the time the discovery suggested a simple mechanical or computer analogy to describe biological systems: 'The discovery of regulator and operator genes, and of repressive regulation of the activity of structural genes, reveals that the genome contains not only a series of blue-prints, but a co-ordinated program of protein synthesis and the means of controlling its execution.' This concept became known as 'genetic programming' (J. Monod and F. Jacob, Genetic regulatory mechanisms in the synthesis of proteins. *J Mol Biol* 1961/3: 318–56).
6. For an extensive discussion of these alternative impacts of the environment during development, see P. D. Gluckman et al., Environmental influences during development and their later consequences for health and disease: implications for the interpretation of empirical studies. *Proc Royal Soc Lond B* 2004/272: 671–7.
7. C. M. Lively, Predator-induced shell dimorphism in the acorn barnacle *Chthamalus anisopoma*. *Evolution* 1986/40: 232–42.
8. See also Ch. 1 n. 35. Although they are most commonly reported in invertebrates, amphibians, and reptiles, such predator-induced responses also occur in mammals. Analysis of records of fur trading by the Hudson's Bay Company revealed a regular ten-year cycle in numbers of the Canadian snowshoe hare and its main predator the lynx. One of the drivers of this cycle may be increased predation pressure at the peak of lynx numbers causing increased stress in the hares and reducing their reproductive output (R. Boonstra et al., The impact of predator-induced stress on the snowshoe hare cycle. *Ecol Monogr* 1998/68: 371–94).
9. R. Yehuda et al., Transgenerational effects of posttraumatic stress disorder in

babies of mothers exposed to the World Trade Center attacks during pregnancy. *J Clin Endocrinol Metab* 2005/90: 4115–18.

10. Professor J. Seckl (Edinburgh), personal communication.
11. D. M. Gardiner and S. V. Bryant, Molecular mechanisms in the control of limb regeneration: the role of homeobox genes. *Int J Dev Biol* 1996/40: 797–805.
12. P. D. Gluckman et al., Environmental influences during development and their later consequences for health and disease: implications for the interpretation of empirical studies. *Proc Royal Soc Lond B* 2004/272: 671–7; P. D. Gluckman and M. A. Hanson, Living with the past: evolution, development and patterns of disease. *Science* 2004/305: 1773–6.
13. P. D. Gluckman et al., Environmental influences during development and their later consequences for health and disease: implications for the interpretation of empirical studies. *Proc Royal Soc Lond* B 2004/272: 671–7; P. D. Gluckman et al., The fetal, neonatal and infant environments: the long-term consequences for disease risk. *Early Hum Dev* 2005/81: 51–9.
14. See n. 13 above.
15. 15% of human mass at birth is fat; most other mammals have less than 5% (C. W. Kuzawa, Adipose tissue in human infancy and childhood: an evolutionary perspective. *Yearbook Physical Anthropol* 1998/41: 177–209).
16. B. Bogin, *Patterns of Human Growth*, 2nd edn. Cambridge: Cambridge University Press, 1999.
17. R. Lewin and R. A. Foley, *Principles of Human Evolution*, 2nd edn. Oxford: Blackwell, 2003.
18. See n. 15 above.
19. For excellent reviews of these alternatives, see: M. F. Small, Mother's little helpers. *New Scientist* 7 Dec. 2002: 44–7; N. Blurton Jones and F. W. Marlowe, Selection for delayed maturity: does it take 20 years to learn to hunt and gather? *Hum Nature* 2002/13: 199–238.
20. In general, evolutionary theory suggests that we age because natural selection does not oppose deleterious mutations or trade-offs acting later in life. Within this general approach, specific models have been proposed. The *mutation accumulation* model of Peter Medawar suggests that there are numerous genes that contribute to senescence—individually, each has only small effects but the accumulation of many detrimental alleles during evolution decreases individual survival. The *antagonistic pleiotropy* model favoured by George Williams invokes pleiotropy, the idea that a single gene may have multiple effects, to propose that an allele which has beneficial effects early in life but detrimental effects later in life will be strongly favoured by selection; alleles with the opposite effect will be selected against. The *disposable soma* model developed by Tom Kirkwood recasts the antagonistic pleiotropy model from the perspective of life-history theory by proposing that high investment in early-life growth and reproduction, which is desirable in terms of Darwinian fitness, will reduce the resources available for

maintenance and repair in later life (M. R. Rose, *Evolutionary Biology of Aging*. New York: Oxford University Press, 1991).

21. The possible exception is the role of the grandmother—see Ch. 9.
22. J. Endler, *Natural Selection in the Wild*. Princeton: Princeton University Press, 1986; J. Weiner, *The Beak of the Finch: A Story of Evolution in our Time*. New York: Alfred A. Knopf, Inc., 1994.
23. P. Beldade and P. M. Brakefield, The genetics and evo-devo of butterfly wing patterns. *Nature Rev Genet* 2002/3: 442–52.
24. T. H. Horton, Fetal origins of developmental plasticity: animal models of induced life history variations. *Am J Hum Biol* 2005/17: 34–43.
25. M. B. Renfree and G. Shaw, Diapause. *Annu Rev Physiol* 2000/62: 353–75. Embryonic diapause has evolved independently at least seven times and is reported in over 100 species of mammal, including over seventy eutherian (true placental) species. Diapause is generally seasonally regulated by changes in melatonin, but then mediated more directly by altered prolactin levels. This is why cessation of suckling can override seasonal effects in the wallaby.
26. Mathematical modelling has been used to explore the conditions under which prediction is advantageous (E. Jablonka et al., The adaptive advantage of phenotypic memory in changing environments. *Phil Trans R Soc Lond B* 1995/350: 133–41; see also N. A. Moran, The evolutionary maintenance of alternative phenotypes. *Am Nat* 1992/139: 971–89; S. E. Sultan and H. G. Spencer, Metapopulation structure favors plasticity over local adaptation. *Am Nat* 2002/160: 271–83). The benefits depend on the accuracy of the prediction, the pattern of environmental change, and the extent of match or mismatch to the new environment—the consequences for survival of being wrong. These need not be symmetrical between states and the environment. Getting the prediction wrong in one direction may have a greater cost than getting it wrong in another. If you predict a sunny day and do not have an umbrella with you when it rains, then the cost of the wrong prediction is greater than predicting a wet day and having an umbrella with you that you did not need to use. But if the umbrella was very heavy and carrying it had a high energetic cost, the relative costs and benefits of being wrong or right would change.
27. See Introduction n. 17.
28. The fetus is entirely dependent on the mother for supply of nutrients. In particular, it needs large amounts of glucose to metabolize for energy production and of amino acids to synthesize proteins for its developing tissues. But these water-soluble molecules cannot easily move across the several layers of lipid-containing cell membranes which separate the fetal and maternal circulations. The problem is solved by specific transport proteins which pump nutrients across the placenta. There are two transporter molecules for glucose, one for transporting glucose from mother to placenta and the other from placenta to fetus, and a whole family of pumps each specializing in a different group of amino acids (C. Sibley et al.,

Placental transporter activity and expression in relation to fetal growth. *Exp Physiol* 1997/82: 389–402). Provision of nutrients to the fetus is controlled by the amount and activity of these transport proteins, and we are now beginning to understand how this process can be affected by various factors. For example, the activity of amino acid transporters is reduced in intrauterine growth restriction. Nicotine is known to have several deleterious effects during pregnancy. Not only does it constrict the mother's blood vessels, resulting in less oxygen being available for the fetus, it also inhibits several of the amino acid transporters in the placenta (T. Regnault et al., Placental amino acid transport systems and fetal growth restriction: a workshop report. *Placenta* 2005/26: S76–S80; A. Pastrakuljic et al., Maternal cocaine use and cigarette smoking in pregnancy in relation to amino acid transport and fetal growth. *Placenta* 1999/20: 499–512).

29. This is an over-simplistic classification implying that there are two extreme strategies. This is not the case—species can combine components of their life history in many ways (S. C. Stearns, *The Evolution of Life Histories*. Oxford: Oxford University Press, 1992).
30. The science of study of these trade-offs is part of life-history theory. For further discussion on life-history method, see S. C. Stearns, *The Evolution of Life Histories*. Oxford: Oxford University Press, 1992; D. A. Roff, *Life History Evolution*. Sunderland, Mass.: Sinauer Associates, 2001.
31. R. G. J. Westendorp and T. B. L. Kirkwood, Human longevity at the cost of reproductive success. *Nature* 1998/396: 743–6. There may be difficulties in the analysis of such a closely related and specialist human population (including the confounding effects of congenital disorders and of the necessity to obtain a male heir) but reanalysis of the same population using a more extensive dataset has confirmed the conclusion (G. Doblhammer and J. Oeppen, Reproduction and longevity among the British peerage: the effect of frailty and health selection. *Proc Biol Sci* 2003/270: 1541–7). A similar study using a German population of mixed socioeconomic status found that poverty increased the strength of the correlation (J. E. Lycett et al., Longevity and the costs of reproduction in a historical human population. *Proc R Soc Lond B* 2000/267: 31–5).
32. Shakespeare tells us in *Macbeth* that MacDuff was 'from his mother's womb untimely ripp'd', and there are other accounts of caesarian sections from the beginning of recorded human history. Initially the procedure was only performed to save the baby in situations where the mother was unlikely to survive the birth, and it is only relatively recently that a caesarian has become safe for mother and baby—even as late as 1865, maternal mortality from the operation in Britain was reported to be 85%. The mother of Julius Caesar is known to have been alive during his adulthood, making it extremely unlikely that he was delivered by the procedure that bears his name; the true origin of the term is unclear, although it has been suggested that it derives from a Roman ('Caesar's') law of the time that made the procedure mandatory.

33. W. Trevathan, Fetal emergence patterns in evolutionary perspective. *Am Anthropol* 1988/90: 674–81.
34. The elucidation of the reproductive physiology of the platypus led to one of the most concise scientific reports in history—the famous telegram 'Monotremes oviparous, ovum meroblastic' sent by William Caldwell from the Australian outback to the British Association meeting of 1884, thus conveying that monotremes lay eggs that contain a large amount of yolk and divide in a way typical of reptiles or birds rather than of mammals.

Chapter 4

1. M. Brunet et al., A new hominid from the Upper Miocene of Chad, Central Africa. *Nature* 2002/418: 145–51.
2. **http://whqlibdoc.who.int/hq/1994/WHO_EHG_94.11_part1.pdf**.
3. More properly termed cerceria.
4. Two separate species build these mounds: *Amitermes meridionalis* and *Amitermes laurensis*, although the latter does not do so in more southern latitudes (**www.qmuseum.qld.gov.au/inquiry/leaflets/leaflet0034.pdf**).
5. See Introduction n. 8 and Ch. 1 n. 1.
6. F. J. Sulloway, Darwin and his finches; evolution of a legend. *J Hist Biol* 1982/15: 1–53.
7. This process of multiple related species evolving to occupy niches in a geographical zone is called adaptive radiation, and is commonly observed when an ancestor species exploits a new habitat. Such a new habitat may arise from environmental change, from the colonization of a new geographical area, or from the ability to use a previously unavailable portion of the existing environment.
8. See Ch. 1 n. 44.
9. D. Christian, *Maps of Time: An Introduction to Big History*. Berkeley and Los Angeles: University of California Press, 2004.
10. J. Kingdon, *Lowly Origins: Where, When and Why our Ancestors First Stood Up*. Princeton: Princeton University Press, 2003.
11. The tracking of the African origin of humans by use of the Y chromosome is described by Spencer Wells (*Journey of Man: A Genetic Odyssey*. Princeton: Princeton University Press, 2002). Mitochondrial DNA has been used in similar studies to follow maternal lineages (reviewed by Brian Sykes in *The Seven Daughters of Eve: The Science that Reveals our Genetic Ancestry*. New York: W. W. Norton, 2001, and see Ch. 7 n. 7). The most recent mitochondrial DNA studies converge with the Y chromosome studies in giving a date of about 50,000 years ago for the human exodus from Africa (M. Ingman et al., Mitochondrial genome variation and the origin of modern humans. *Nature* 2000/408: 708–13).
12. T. Zerjal et al., The genetic legacy of the Mongols. *Am J Hum Genet* 2003/72:

717–21. Mitochondrial DNA tracking has also been used for specific populations, for instance a recent study of the Ashkenazi (central and eastern European) Jewish population has shown that one-half of the population can be traced back to only four women who lived within the last 2,000 years (D. M. Behar et al., The matrilineal ancestry of Ashkenazi Jewry: portrait of a recent founder event. *Am J Hum Genet* 2006/78: 487–97).

13. Some anthropologists argue for earlier and multiple migrations from Asia into North America. Genetic analysis, the presence in the Americas of three language groups, and the archaeological remains that now point to a human presence in Central and South America by 14,500 to 13,000 years ago have led some (T. G. Schurr and T. G. Sherry, Mitochondrial DNA and Y chromosome diversity and the peopling of the Americas: evolutionary and demographic evidence. *Am J Hum Biol* 2004/16: 420–39) to suggest three separate expansions, of which the first would have been an initial migration from Siberia between 20,000 and 15,000 years ago.
14. See Luigi Cavalli-Sforza's book *Genes, Peoples and Languages* (Berkeley and Los Angeles: University of California Press, 2000).
15. R. Klein and B. Edgar, *The Dawn of Human Culture*. New York: John Wiley & Sons, 2002.
16. S. Mcbrearty and A. S. Brooks, The revolution that wasn't: a new interpretation of the origin of modern human behavior. *J Hum Evol* 2000/39: 453–563.
17. For recent reviews on the theories of the origins of religion see R. Dunbar, We believe. *New Scientist* 28 Jan. 2006: 30–3 and D. S. Wilson, *Darwin's Cathedral: Evolution, Religion and the Nature of Society*. Chicago: University of Chicago Press, 2002.
18. J. Tooby and L. Cosmides, The past explains the present: emotional adaptations and the structure of ancestral environments. *Ethol Sociobiol* 1990/11: 375–424.
19. R. A. Foley, The adaptive legacy of human evolution: a search for the environment of evolutionary adaptedness. *Evol Anthropol* 1996/4: 194–203.
20. We will probably never know exactly what humans ate 50,000 years ago, but a number of tools are available to anthropologists studying the hominid diet. They can seek parallels with the dietary patterns of modern hunter-gatherer societies. They can study the evolution of human jaws and teeth. They can compare the modern human gut with that of other animals to see whether we are structurally most like herbivores or carnivores. They can study fossils directly, particularly by analysing trace elements in tooth enamel and bone—for example, meat is a major source of zinc so carnivore bone is relatively high in that element. A particularly elegant method involves measurement of carbon isotope ratios in tooth and bone extracts. Plants perform photosynthesis in different ways, of which the so-called C_3 pathway is used by woody plants such as trees and bushes and the C_4 pathway is used mostly by grasses. The intricacies of those biochemical pathways result in C_4 plants having a relatively high content of the stable isotope carbon-13. That

distinctive ratio is reflected in the tissues of the animals that eat the plants, so that, using an African example, the carbon isotope ratio in giraffes, which browse on the leaves of trees and bushes, is different from that in zebras, which graze on the grasses. And in turn the tissues of predators higher up the food chain reveal what they have been eating (C. R. Peters and J. C. Vogel, Africa's wild C4 plants and possible early hominid diets. *J Hum Evol* 2005/48: 219–36). It seems likely that before the introduction of agriculture about 10,000 years ago the human diet featured a high proportion of animal protein with relatively little fat and virtually no refined carbohydrates—a clear mismatch with the modern western diet (L. Cordain et al., Origins and evolution of the Western diet: health implications for the 21st century. *Am J Clin Nutr* 2005/81: 341–54).

21. B. Bogin, *Patterns of Human Growth*, 2nd edn. Cambridge: Cambridge University Press, 1999; R. H. Steckel and J. C. Rose (eds.), *The Backbone of History: Health and Nutrition in the Western Hemisphere*. Cambridge: Cambridge University Press, 2002.
22. D. Christian, *Maps of Time: An Introduction to Big History*. Berkeley and Los Angeles: University of California Press, 2004.
23. Although they did develop a form of eel farming using complex weirs.
24. It is thought that climate changes in Mesopotamia between 10,000 and 4,000 years ago were a further impetus to agricultural invention in the fertile crescent. During that period there was an 8% increase in sunlight in the northern hemisphere due to a change in the Earth's axis and this was associated with a 30% increase in rainfall in the fertile crescent. A sunnier, moister climate would favour agriculture. Incidentally, about 4,000 years ago there were further cyclic changes that left the region drier, agriculture became more difficult to sustain, and the archaeological evidence shows an increase in the use of irrigation, shifts in population and to trading as the basis of sustenance—all as responses to this change (T. Flannery, *The Weather Makers: The History and Future Impact of Climate Change*. Sydney: Text Publishing, 2005).
25. D. Christian, *Maps of Time: An Introduction to Big History*. Berkeley and Los Angeles: University of California Press, 2004, p. 196.
26. D. Christian, *Maps of Time: An Introduction to Big History*. Berkeley and Los Angeles: University of California Press, 2004, p. 233.
27. S. C. Stearns, Introducing evolutionary thinking. In S. C. Stearns (ed.), *Evolution in Health and Disease*. New York: Oxford University Press, 1999.
28. The 'hygiene hypothesis' was proposed in the 1980s to explain the increase in childhood allergic disease in the developed world together with the epidemiological observations that exposure to apparently less clean conditions, such as keeping a pet or living on a farm, was protective against such disease. The hypothesis initially suggested that the cleanliness of modern living conditions reduces childhood contact with pathogenic micro-organisms and resets the immune system in a way that increases the likelihood of allergy (specifically,

a shift from a so-called Th1 profile to a Th2 profile, named after the immune cells (T helper, or Th, cells) that initiate these responses). However, the developed world is also experiencing a marked increase in the incidence of autoimmune disorders and inflammatory bowel disease, which paradoxically involve an abnormally increased Th1 response. Consequently, the hygiene hypothesis has recently been restated to suggest that reduced contact with micro-organisms changes the balance of effector and regulatory cells in the immune system, possibly by involvement of the receptors in intestinal cells that sense specific components of the intestinal flora of micro-organisms, and that such imbalance results in autoimmune (Th1-mediated) or allergic (Th2-mediated) disease depending on an individual's predisposition (G. A. W. Rook and L. R. Brunet, Microbes, immunoregulation, and the gut. *Gut* 2005/54: 317–20).

29. G. Davies, *A History of Money from Ancient Times to the Present Day*, 3rd edn. Cardiff: University of Wales Press, 2002.
30. R. Laurence, Writing the Roman metropolis. In H. M. Parkins (ed.), *Roman Urbanisation: Beyond the Consumer City*. London: Routledge, 1997.
31. J. Carcopino and J. H. T. Rowell, *Daily Life in Ancient Rome: The People and the City at the Height of the Empire*, 2nd edn. New Haven: Yale University Press, 2003.
32. *The Twelve Caesars*, by Suetonius Tranquillus. Robert Graves used Suetonius as a source for *I, Claudius* and later published a complete translation, recently reissued (London: Penguin Classics, 2003).
33. Neanderthal burials include elderly individuals with skeletal disease or healed fractures suggesting that, at least in some instances, this culture took care of the sick and injured.
34. W. C. Jordan, *Europe in the High Middle Ages*. London: Penguin Books, 2004.
35. There is an important caveat in interpreting figures about life expectancy. The calculation of average life expectancy *at birth* is largely influenced by childhood mortality. Many live-born infants die in the first five years of life, particularly in the first year, and especially in the first month. In Palaeolithic times infant and child mortality is thought to have exceeded 50%. Therefore a reduction in child mortality has a major impact on population life expectancy. Thus, only as childhood mortality falls do events in later life start having a major effect on average life expectancy. In recorded history there have always been a significant number of individuals living well into their fifth and sixth decades despite this low average life expectancy.
36. From Samuel Johnston's essay series *The Rambler*, entry for 30 Nov. 1751. Now most easily available from sources such as the Electronic Text Center of the University of Virginia Library, **http://etext.lib.virginia.edu/toc/modeng/public/Joh4Ram.html**.
37. The most convenient modern form is electronic text available from, for example, the Voltaire Foundation (**www.voltaire.ox.ac.uk**).

38. The source of the quake was the Gorringe Bank some hundred of kilometres off the coast of Portugal. It was probably the first well-documented earthquake due to the actions of the Prime Minister Sebastio Jose de Carvalho e Melo who sent a questionnaire on its effects to every parish in Portugal (B. McGuire, Swept away. *New Scientist* 22 Oct. 2005: 38–42).
39. R. W. Fogel, *The Escape from Hunger and Premature Death, 1700–2100: Europe, America and the Third World*. Cambridge: Cambridge University Press, 2004.
40. H. de Beer, Observations on the history of Dutch physical stature from the late-Middle Ages to the present. *Econ Hum Biol* 2004/2: 45–55.
41. In the first decade of the 20th century, the anthropologist Franz Boas measured the physical proportions of nearly 18,000 individuals—immigrants to the USA and their children—in a study of human response to a new environment. The results, showing marked differences between immigrant parents and their US-born offspring, provided strong evidence against the then popular doctrine of a biological basis for racial differences (F. Boas, *Changes in Bodily Form of Descendants of Immigrants*. Washington: Government Printing Office, 1911).
42. B. Bogin, *Patterns of Human Growth*, 2nd edn. Cambridge: Cambridge University Press, 1999.
43. D. Cha, *Hmong American Concepts of Health, Healing, and Conventional Medicine*. New York: Routledge, 2003.
44. M. A. Huffman, Animal self-medication and ethno-medicine: exploration and exploitation of the medicinal properties of plants. *Proc Nutr Soc* 2003/62: 371–81.
45. For the historical background to Ross's discovery see J. Guillemin, Choosing scientific patrimony: Sir Ronald Ross, Alphonse Laveran, and the mosquito-vector hypothesis for malaria. *J Hist Med Allied Sci* 2002/57: 385–409. For current information on malaria and efforts to control it, see **www.who.int/mediacentre/factsheets/fs094/en/**.
46. The development of the contraceptive pill was based on innovations in chemistry and in medicine. The chemistry began in the 1940s with the discovery by Dr Russell Marker (who also invented the octane rating system for hydrocarbon fuel) that Mexican yams contain large amounts of diosgenin, a convenient precursor molecule for the synthesis of steroid hormones. The medical story began in the early 1950s, when the American birth control activist Margaret Sanger encouraged and arranged funding for Dr Gregory Pincus to develop an effective oral contraceptive. The first trials of the pill took place in Puerto Rico in the mid-1950s, it was approved by the US regulatory authorities in 1960, and registration in most other countries followed quickly. But it was not until 1999 that the contraceptive pill was approved in Japan.
47. **www.statistics.gov.uk/cci/nugget.asp?id=951**.
48. **http://news.bbc.co.uk/1/hi/world/europe/4199839.stm**.
49. A. Gibbons, Solving the brain's energy crisis. *Science* 1998/280: 1345–7.
50. The observations of anthropologist Richard Lee that the requirements of the

Dobe !Kung could be satisfied by little more than two days of work each week prompted Marshall Sahlins to contrast the ability of hunter-gatherer societies to achieve affluence by desiring little with the unlimited material desires of modern industrial society (R. B. Lee, What hunters do for a living, or, how to make out on scarce resources; M. Sahlins, Notes on the original affluent society. In R. B. Lee and I. Devore (eds.), *Man the Hunter*. New York: Aldine Publishing Co., 1968).

51. Dunbar has extrapolated from the mathematical relationship that exists between brain size and social group size across primates to calculate human clan size in the Palaeolithic (R. Dunbar, Coevolution of neocortical size, group size and language in humans. *Behav Brain Sci* 1993/16: 681–735). For a discussion of human groups from the perspective of evolutionary psychology, see R. A. Foley, Evolutionary perspectives on the origins of human social institutions. *Proc Br Acad* 2001/110: 171–95.
52. Global warming is unequivocally the result of human impact on the environment (see the report from the US National Research Council *Climate Change Science: An Analysis of Some Key Questions*; **www.nap.edu/catalog/10139.html?onpi_newsdoc06062001**)—denial will not address the issues. Careful calculation of carbon dioxide levels from ice-core analysis shows that levels started to rise about 200 years ago, coincident with the industrial revolution. Methane levels started to rise about 8,000 years ago—perhaps the first evidence of the impact of humans on the planetary environment and probably a result of the development of agriculture in eastern Asia and the domestication of rice, because paddy fields are a major source of methane (T. Flannery, *The Weather Makers: The History and Future Impact of Climate Change*. Sydney: Text Publishing, 2005).

Chapter 5

1. But much of our body works on a very short time scale. Our biological clocks are entrained by a part of the brain called the hypothalamus which ensures that our daily rhythm of hormone secretion and body functions such as sleeping is coordinated with night and day, because we are a species designed to largely function by day and sleep by night. Our rhythms are very tightly controlled. For example, the secretion of our primary stress hormone, cortisol, by the adrenal gland is timed for the early morning to set our body activities for the day, our secretion of a fat mobilizing hormone, growth hormone, occurs at the beginning of our first deep sleep cycle and this mobilizes stored fuels to allow us to fast overnight. Our body temperature has a fundamental 24-hour rhythm, and so does our heart rate and many other aspects of our physiology.

   The two authors know a lot about jet lag. It happens because we shift our position in the light cycle very rapidly by flying across twelve time zones from Auckland to London or vice versa in twenty-six hours. The body does not have time to readjust its settings to synchronize its various biological rhythms to the light/dark cycle and so we find ourselves mismatched in what our body wants to do and when it should do it—we can't sleep at the right time, our urine output is greater in the evening than in the morning, we are hungry at the wrong times; it is a very acute but fortunately transient form of mismatch. Internal regulation allows us to adapt to minor transient changes within our environment.

2. For a splendid discussion on why there are constraints on selection and why these constraints can give a false impression that there is a directionality to selection see S. J. Gould *Life's Grandeur: The Spread of Excellence from Plato to Darwin*. London: Jonathan Cape, 1996.
3. For a recent discussion see M. W. Kirschner and J. Gerhart, *The Plausibility of Life: Resolving Darwin's Dilemma*. New Haven: Yale University Press, 2005.
4. There is increasing molecular evidence that some degree of selection has continued in humans since the upper Palaeolithic. A recent paper (B. E. Voight et al., A map of recent positive selection in the human genome. *PLoS Biology* 2006/4: 446–58) reports scanning the human genome for single nucleotide polymorphisms in three separate human populations from Africa, Europe, and East Asia. They analysed several hundred thousand polymorphisms and looked for differences within and between populations. They identified genes that they concluded showed signs of recent (within 10,000 years) selection, including genes associated with nutrition and metabolism (among them the lactase gene) and a number associated with reproduction and fertility. Other studies (P. D. Evans et al., Microcephalin, a gene regulating brain size, continues to evolve adaptively in humans. *Science* 2005/309: 1717–20; N. Mekel-Bobrov et al., Ongoing adaptive evolution of APSM, a brain size determinant in *Homo sapiens*. *Science* 2005/309: 1720–2) suggest that genes regulating brain size have undergone positive selection in the last 40,000 and 10,000 years respectively. These are molecular studies and the selection is inferred from statistical techniques. The adaptive correlates of most of these changes in allele frequency are speculative and it may be that sexual selection has played a significant role in some of these frequency changes.
5. The issue of the importance of ongoing human evolution by Darwinian mechanisms remains contentious. On the one hand, for example, Christopher Wills argues in his book *Children of Prometheus: The Accelerating Pace of Human Evolution* (New York: Perseus Books, 1998) that the human willingness to explore extremes, both of environment and of experience, and the consequences of our unprecedented ability to alter our surroundings are leading to an accelerating rate of evolutionary change with each human generation. The molecular evidence for selection continuing since the upper Palaeolithic is reviewed in n. 4 above. On the other hand, some evolutionary biologists, such as Steve Jones, have argued that natural selection is of much lesser importance for the modern human in determining our destiny as a species (**www.open2.net/truthwillout/evolution/article/evolution_jones.htm**). Our view, as detailed in the text, is that while

selection was important in the major transitions of agriculture and settlement and in responses to infection, it is unlikely to play a significant role in providing solutions to the mismatches we now face. For a recent review of the debate, see K. Douglas, Are we still evolving? *New Scientist* 11 Mar. 2006: 30–3.

6. D. H. Hubel and T. N. Wiesel, *Brain and Visual Perception: The Story of a 25-Year Collaboration*. New York: Oxford University Press, 2004.
7. For a more extensive discussion of plasticity and its costs see S. Nylin and K. Gotthard, Plasticity in life history traits. *Annu Rev Entomol* 1998/43: 63–83. To quote: 'Plasticity seems to provide a wonderful way of adapting the phenotype; so why are not all organisms perfectly plastic? Is the main reason that plasticity is hard to evolve, or is it that it disappears quickly in stable environments? If the latter, are there costs to plasticity, or is it the time lags between induction and expression that select against (persistently) plastic phenotypes?' An even more comprehensive review that gives strong impetus to environmental factors operative in development is found in a major book by Mary-Jane West-Eberhard (*Developmental Plasticity and Evolution*. New York: Oxford University Press, 2003).
8. The fetus has a very limited capacity to sense the outside world directly; for example, in late gestation the human fetus can respond to vibratory and acoustic signals but even these are modulated by the passage of the sound through the mother, and electrophysiological studies in experimental animals and premature infants show that while brainstem responses may be present in mid-gestation the cortical awareness of these signals probably develops very late (C. J. Cook et al., Brainstem auditory evoked potentials in the fetal lamb, in utero. *J Dev Physiol* 1987/9: 429–40; L. A. Werner, Early development of the human auditory system. In R. A. Polin et al. (eds.), *Fetal and Neonatal Physiology*, vol. ii, 3rd edn. Philadelphia: W. B. Saunders, 2004).
9. For a diagrammatic representation of the very big difference in relative pelvic capacity see W. R. Trevathan, Evolutionary obstetrics. In W. R. Trevathan et al. (eds.), *Evolutionary Medicine*. New York: Oxford University Press, 1999.
10. P. D. Gluckman and M. A. Hanson, Maternal constraint of fetal growth and its consequences. *Semin Fetal Neonatal Med* 2004/9: 419–25; P. D. Gluckman et al., The fetal, neonatal and infant environments—the long-term consequences for disease risk. *Early Hum Dev* 2005/81: 51–9.
11. For more extensive discussion of maternal constraint see P. D. Gluckman and M. A. Hanson, *The Fetal Matrix: Evolution, Development, and Disease*. Cambridge: Cambridge University Press, 2005.
12. J. B. Birdsell, Ecological influences on Australian aboriginal social organization. In I. S. Berstein and E. O. Smith (eds.), *Primate Ecology and Human Origins: Ecological Influences and Social Organization*. New York: Garland, 1979.
13. In some cultures such as the Australian Aborigine, infanticide was also used to ensure an adequate birth interval because the mother could not adequately support two pre-weaned infants in a nomadic way of life (reference in n. 12 above).

Not only is breast feeding the optimal means of infant nutrition in early life but it also suppresses mother's menstruation and provides initially a high degree of contraception. The resulting increase in birth spacing promotes infant survival, because the mother has time to provide adequate care, and allows her to recuperate more fully between pregnancies (see **www.unsystem.org/scn/archives/npp11/ch05.htm**).

Chapter 6

1. Slightly paraphrased from R. Sapolsky, *A Primate's Memoir: Love, Death and Baboons in East Africa*. London: Jonathan Cape, 2001.
2. M. Abley, *Spoken Here: Travels among Threatened Languages*. New York: Houghton Mifflin, 2003.
3. See Ch. 5 n. 12.
4. G. Weisfeld, *Evolutionary Principles of Human Adolescence*. New York: Basic Books, 1999; A. Schlegel and H. Barry, The evolutionary significance of adolescent initiation ceremonies. *Am Ethol* 1980/7: 696–715.
5. S. Humphries and D. J. Stevens, Reproductive biology: out with a bang. *Nature* 410: 758–9. And this mouse is not unique. A larger kitten-sized ferret-like marsupial, the northern quoll, *Dasyurus hallucatus*, which lives in northern Australia has a very similar behaviour (M. Oakwood et al., Semelparity in a large marsupial. *Proc Biol Sci* 2001/268: 407–11).
6. B. Bogin, *Patterns of Human Growth*, 2nd edn. Cambridge: Cambridge University Press, 1999.
7. R. Dunbar, *The Human Story: A New History of Mankind's Evolution*. London: Faber & Faber, 2004.
8. J. Mercader et al., Excavation of a chimpanzee stone tool site in the African rainforest. *Science* 2002/296: 1452–5.
9. J. Altman et al., Determinants of reproductive success in savannah baboons, *Papio cynocephalus*. In T. H. Clutton Brock (ed.), *Reproductive Success*. Chicago: University of Chicago Press, 1988; A. H. Harcourt et al., Demography of *Gorilla gorilla*. *J Zool* 1981/195: 215–33. There is not much reason to think our Palaeolithic and Neolithic ancestors fared much better, and indeed the archaeological evidence suggests that perhaps 50% or more of children of our hunter-gatherer ancestors died before the age of 5 years. Childhood mortality in modern hunter-gatherers is not so very different (N. Howell, *Demography of the Dobe !Kung*. New York: Academic Press, 1979).
10. M. F. Small, Mother's little helpers. *New Scientist* 7 Dec. 2002: 44–7.
11. H. Kaplan et al., A theory of human life history evolution: diet, intelligence and longevity. *Evol Anthropol* 2002/9: 149–86.
12. N. Blurton Jones and F. W. Marlowe, Selection for delayed maturity: does it take 20 years to learn to hunt and gather? *Hum Nature* 2002/13: 199–238.

13. R. Bliege Bird and D. W. Bird, Constraints of knowing or constraints of growing? Fishing and collecting by the people of Mer. *Hum Nat* 2002/13: 239–67.
14. Coping with social complexity requires what psychologists call 'Theory of Mind'—the ability to recognize that other people's view of the world might be different from one's own and to deal with the increasing 'levels of intentionality' implicit in statements such as 'I believe that Peter thinks that Mark knows . . .' Human children begin to develop Theory of Mind at about the age of 4 years, and there is evidence for a limited Theory of Mind in higher primates, particularly chimpanzees (R. Dunbar, *Grooming, Gossip and the Evolution of Language*. London: Faber & Faber, 1996).
15. R. A. Foley, Evolutionary perspectives on the origins of human social institutions. *Proc Br Acad* 2001/110: 171–95. R. Dunbar, *Grooming, Gossip and the Evolution of Language*. London: Faber & Faber, 1996.
16. Not to be confused with melanin, which is the black compound responsible for the pigmentation of skin and hair. Because melatonin levels are higher at night than in the day and melatonin is used by the body to signal that it is dark, melatonin has been used (with variable reports as to its efficacy) in the management of jet lag, to try and shift the body clock more rapidly to match a new time zone.
17. P. D. Gluckman et al., The human fetal hypothalamus and pituitary gland: the maturation of neuroendocrine mechanisms controlling the secretion of fetal pituitary growth hormone, prolactin, gonadotropin and adrenocorticotropin related hormones. In K. Ryan and D. Tulchinsky (eds.) *Maternal-Fetal Endocrinology*. Philadelphia: Saunders, 1980.
18. See **http://wiseli.engr.wisc.edu/news/Summers.htm**.
19. Proponents of an organic basis for human sexual orientation draw on a number of supporting observations. First, there is modest evidence for a genetic contribution to sexual orientation, with data for maternal-line transmission suggesting X-linkage. Secondly, although the fraternal birth order effect—the well-established finding that homosexual men have more elder brothers than do heterosexual men—can be interpreted as an outcome of environment, some investigators have proposed that a succession of male fetuses causes the maternal immune system to produce male-specific antibodies that in some way affect brain development and sexual preference in later male offspring. Thirdly, and perhaps more plausibly, given the overwhelming evidence for the role of testosterone (or its absence) in the fetal and neonatal periods in determining physical sexual development and hypothalamic function in mammals, it seems not unlikely that this hormone will also affect psychosexual development (D. F. Swaab, The role of hypothalamus and endocrine system in sexuality. In J. S. Hyde (ed.), *Biological Substrates of Human Sexuality*. Washington: American Psychological Association, 2005). The relationship between hormone levels in the fetal brain and sexual orientation twenty years later is clearly rather difficult to study in humans, but

markers of prenatal androgen exposure do exist and they do appear to correlate, albeit weakly, with sexual orientation. One of these markers is the ratio of the lengths of the second and fourth fingers. This ratio is dimorphic in humans—lower in males than in females—but shows differences between homosexual and heterosexual women, and less clearly between homosexual and heterosexual men, that in turn suggest differences in early androgen exposure. Finally, there is some evidence from human autopsy studies, and broadly consistent evidence from animal studies, of differences in brain structure that correlate with sexual orientation—and the differences in the animal model could be linked with alterations in steroid hormone production (Q. Rahman, The neurodevelopment of human sexual orientation. *Neurosci Biobehav Rev* 2005/29: 1057–66). This is not the book to consider this fascinating topic in further detail, and we leave you measuring your fingers.

20. B. S. McEwen, Gonadal steroid influences on brain development and sexual differentiation. *Int Rev Physiol.* 1983/27: 99–145; I. J. Clarke et al., Effects of testosterone implants in pregnant ewes on their female offspring. *J Embryol Exp Morphol* 1976/36: 87–99.
21. When ovulation occurs, the egg is released from its nurturing follicle in the ovary and starts its journey down the Fallopian tube to the uterus. The empty follicle in the ovary starts making a second hormone, progesterone, until it is reabsorbed. This in turn affects the control system and changes the release of gonadotropins, and thus in the absence of pregnancy this process becomes cyclical—menstruation occurring each time progesterone secretion falls. Menstruation is the shedding of the uterine lining which has grown under stimulation by progesterone to be ready to receive the embryo if pregnancy occurs. In the absence of pregnancy the fall in progesterone leads to the lining being shed. In pregnancy the fertilized egg signals the ovary to keep making progesterone until the primitive placenta can take over the production of this hormone. The high progesterone levels stop any further ovulation while the woman is pregnant. In the absence of pregnancy the woman will continue to ovulate on a monthly cycle until the menopause.
22. The anthropologist Barry Bogin from Michigan has argued (B. Bogin, *Patterns of Human Growth*, 2nd edn. Cambridge: Cambridge University Press, 1999) that the pubertal growth spurt is about catching up. He has suggested that humans delay growth in childhood to invest their available resources initially for energetically demanding brain growth, and only when that is over can they afford to invest in linear growth. The argument is not entirely convincing—for example, other large primates have relatively large brains yet with the exception of one or two observations in captive male chimps and gorillas there is no hint of a linear growth spurt at all in our closest relatives. The timing of brain growth is about the same in both sexes yet males have a growth spurt delayed by one to two years, which is also difficult to reconcile with this theory. In the female a role for natural

selection in generating the pubertal growth spurt seems probable. Women who grew to be taller at the time of reproductive competence were more likely to survive the risky process of childbirth and would have larger babies who might survive infancy better. Sexual selection could reinforce this if males perceived that larger females were more likely to be successful mates (P. D. Gluckman and M. A. Hanson, Evolution, development and timing of puberty. *Trends Endocrinol Metab* 2006/17: 7–12). In the male the pubertal growth spurt could well have been influenced by sexual selection. It evolved rapidly and distinctly. While some adaptive advantage can be attributed to the growth spurt, the observation that it has not evolved in any other species suggests that any such advantage is of lesser importance. Just as humans probably lost the bulk of their body hair through sexual selection, males were probably selected for sexual success based on their body size. This might be either through sexual selection driven by female choice or through the male's greater physical dominance (sexual competition). One further consideration is that the pattern of the pubertal growth spurt may have been influenced by the relative roles of fertile males and females in Palaeolithic society. In traditional populations, juvenile male children are immune from inter-male conflict (generally over power and mating rights) but sexually competent males are not. Therefore, there is an advantage in males having a later growth spurt to delay being caught up in conflict and then growing rapidly once sexually competent to be able to protect themselves. Females on the other hand grow to be physically mature prior to being reproductively competent. This ensures that there is a chance for their first child to be adequately nourished *in utero* and thus to survive. Marriage or mating would have been an early event in females (and still is in traditional societies such the Masai and the Aborigine) to maximize their lifetime reproductive fitness. A feature of many hunter-gatherers is that generally females disperse to other clans to breed whereas males stay with the clan in which they were born. In such societies it would be advantageous for females to look mature before being reproductively competent. They can then move to another tribe as a 'bride' before they could possibly be pregnant; if they could reproduce before they were dispersed the taboos against intra-kin sexuality might easily be broken. (We acknowledge the invaluable discussion with Rob Foley and Marta Lahr from Cambridge in developing this argument.)

23. S. de Muinck Keizer-Schrama and D. Mul, Trends in pubertal development in Europe. *Hum Reprod Update* 2001/7: 287–91.
24. P. D. Gluckman and M. A. Hanson, Evolution, development and timing of puberty. *Trends Endocrinol Metab* 2006/17: 7–12.
25. C. Cooper et al., Childhood growth and age at menarche. *Br J Obstet Gynaecol* 1996/103: 814–17.
26. A.-S. Parent et al., The timing of normal puberty and the age limits of sexual precocity: variations around the world, secular trends, and changes after migration. *Endocr Rev* 2003/244: 668–93.

27. The best-documented case is the exceptional situation in Puerto Rico, where high rates of early breast development in young girls without other signs of puberty (premature thelarche) have been recorded since 1979. Environmental contamination with oestrogens originating from food additives or pharmaceutical waste products have been suggested but the evidence is not strong. One controversial study has shown the presence of phthalate esters, which are used as plasticizers in food packaging and have some, although low, oestrogenic activity, at significantly higher levels in affected girls than in unaffected girls of the same age (I. Colon et al., Identification of phthalate esters in the serum of young Puerto Rican girls with premature breast development. *Environ Health Perspect* 2000/108: 895–900). Nevertheless, the phenomenon remains unexplained.
28. J. T. Smith and B. J. Waddell, Increased fetal glucocorticoid exposure delays puberty in postnatal life. *Endocrinology* 2000/141: 2442–8.
29. J. Giedd, Structural magnetic resonance imaging of the adolescent brain. *Ann NY Acad Sci* 2004/1021: 77–85.
30. Anthropologists have also tried to use calculations of height from skeletal remains to assess historical trends in health. One study looked at the trends in height from skeletal remains of native Americans over the last 8,000 years and suggests that these people were relatively tall while they were solely hunter-gatherers, but that their heights started to fall as agriculture and settlement started to appear. A nadir was reached in the 18th century, and then, slowly, height started to rise again (B. Bogin and R. Keep, Eight thousand years of economic and political history in Latin America revealed by anthropometry. *Ann Hum Biol* 1999/26: 333–51). Adult height has increased much more rapidly recently as conditions have improved. This study has the limitations of only isolated data points in the early years, but taken together with other studies it does suggest that growth is related to the general conditions in which we live, and have lived, and that hunter-gatherers could grow to heights equivalent to those reached in modern times. This is shown most dramatically for the Maya, who were so stunted in their native environments that researchers in the recent past thought that they were an American form of pygmy. But when some of them moved from their subsistence environment to Florida, they showed rapid linear growth in one generation (B. Bogin, *Patterns of Human Growth*, 2nd edn. Cambridge: Cambridge University Press, 1999). The height of American indigenous people is now back to where it was some 8,000 years ago—so much for human progress.
31. The original description of the secular decline in the age of menarche was by (J. M. Tanner, *Growth at Adolescence*. Oxford: Blackwell Scientific Publications, 1962) who reported a faster decline, but the population estimates from the 1860s were based on small numbers which may have not been fully representative. Nevertheless, there is now a considerable literature confirming his basic conclusions about the magnitude of the trend, at least in European populations (A.-S. Parent et al., The timing of normal puberty and the age limits of sexual precocity:

variations around the world, secular trends, and changes after migration. *Endocr Rev* 2003/244: 668–93; J. E. Brudevoll et al., Menarchal age in Oslo during the last 140 years. *Ann Hum Biol* 1979/6: 407–16).

32. P. B. Eveleth, and J. M. Tanner *Worldwide Variation in Human Growth*, 2nd edn. Cambridge: Cambridge University Press, 1990.
33. It may even be that the age of puberty could fall below its evolutionarily determined value. Our colleague Sir Patrick Bateson from Cambridge has suggested that perhaps the excessive obesity now observed in childhood could precipitate a further reduction in the age of puberty by signalling particular bounty.
34. The 'normal' lowest age for onset of puberty in girls was recently revised to 7 years for white girls and 6 years for black girls. Onset of puberty in boys before 9 years is considered precocious (P. B. Kaplowitz and S. E. Oberfield. Reexamination of the age limit for defining when puberty is precocious in girls in the United States: implications for evaluation and treatment. *Pediatrics* 1999/104: 936–41).

Chapter 7

1. The Cathars, or Albigensians, were an ascetic religious group who appeared in Europe in the 10th century and prospered in the liberal atmosphere of the area of southern France known as the Languedoc during the 12th and early 13th centuries. There is debate over whether Catharism should be considered a Christian heresy or a separate religion, but their beliefs brought them into conflict with the Catholic Church, and after peaceful attempts to convert the Cathars had failed Pope Innocent III proclaimed the Albigensian Crusade in 1209. The resulting war and the subsequent inquisition, driven both by religious fervour and by the territorial claims of northern France, resulted in the deaths of half a million people—the last execution of a Cathar occurred in 1321. Some historians see the success of this internal European crusade, in contrast to the dubious achievements of the crusades in the Middle East, as a grim precedent for later European religious wars and 'ethnic cleansing'.
2. **www.geocities.com/triple-moon/articles/venusfig.html**.
3. **www.newstarget.com/002173.html**.
4. **www.post-gazette.com/pg/05072/470035.stm**.
5. **www.guardian.co.uk/australia/story/0,,1725977,00.html**.
6. Childhood obesity is an acute health crisis, according to the World Health Organization. Globally, 10% of schoolchildren between 5 and 17 years old are overweight or obese; (**www.who.int/mediacentre/news/releases/2004/pr81/en/**). The prevalence of childhood obesity appears to be increasing rapidly. Different studies have looked at different age groups and used different definitions of obesity, so global comparisons are difficult. In the USA, 25% of children aged 10 to 16 years are affected (I. Janssen et al., Comparison of overweight and obesity prevalence in school-aged youth from 34 countries and their relationships with

physical activity and dietary patterns. *Obes Rev* 2005/6: 123–32). In Europe, the prevalence of overweight and obesity in children aged 7 to 11 years ranges from 10% in the Russian Federation to 35% in Italy (T. Lobstein and M.-L. Frelut, Prevalence of overweight among children in Europe. *Obes Rev* 2003/4: 195–200). Over 20% of Australian children are overweight (J. McLennan, Obesity in children: tackling a growing problem. *Aust Fam Physician* 2004/33: 33–6). But childhood obesity is not limited to the developed countries. Overweight was found in 18% of Indian children and adolescents in a sample from New Delhi and in 10% of Pakistani 8- to 12-year-olds in Karachi (A. Misra et al., High prevalence of obesity and associated risk factors in urban children in India and Pakistan. *Diabet Res Clin Pract* 2006/71: 101–2). Rates of 12–13% have been reported for urban children in China and Brazil (**www.who.int/mediacentre/news/releases/2004/pr81/en/**). Childhood obesity is related to socioeconomic status in both developed and developing countries—but the direction of the relationship is opposite for the two situations. In fully westernized countries, children from the more disadvantaged social groups are more likely to be overweight. In the developing world, obesity is a problem of wealthier urban children in areas that have already begun economic development (J. J. Reilly, Descriptive epidemiology and health consequences of childhood obesity. *Best Pract Res Clin Endocrinol Metab* 2005/19: 327–41). The rate of change in prevalence has been alarmingly rapid: from 8% in 1984 to 20% in 1998 in the UK, from 3% in 1960 to 16% in 2000 in France, from 23% in 1986 to 35% in 1996 in Spain, from 15% in the 1970s to 25% in the 1990s in the USA, and from 4% in the mid-1970s to 13% in 1997 in Brazil (sources above and **www.euro.who.int/document/mediacentre/fs1305e.pdf**). It is difficult not to attribute these alarming increases in part to a toxic environment of aggressively marketed food products of poor nutritional quality coupled with an increasingly sedentary lifestyle (C. B. Ebbeling et al., Childhood obesity: public-health crisis, common sense cure. *Lancet* 2002/360: 473–82).

7. Mitochondria themselves have a fascinating evolutionary history. They are ancient bacteria which got incorporated into single-cell organisms billions of years ago and started using the cell's machinery to aid their reproduction. This symbiosis morphed into a permanent arrangement by which mitochondria became a permanent part of the cell's machinery. Because of this history, mitochondria have their own mini-chromosome that carries the genes for mitochondrial enzymes involved in energy metabolism. But sperm have no mitochondria and eggs have many. Thus, all the mitochondria in every cell in our body come from our mother. Hence mitochondrial DNA is only inherited from our mother, and it is this knowledge that is used in some modern genomic studies of human ancestry.
8. S. Weidensaul, *Living on the Wind: Across the Hemisphere with Migratory Birds*. New York: North Point Press, 1999.

9. A. C. Bell et al., The road to obesity or the path to prevention: motorized transportation and obesity in China. *Obes Res* 2002/10: 277–83.
10. R. J. Hancox et al., Association between child and adolescent television viewing and adult health: a longitudinal birth cohort study. *Lancet* 2004/364: 257–62; R. M. Viner and T. J. Cole, Television viewing in early childhood predicts adult body mass index. *J Pediatr* 2005/147: 429–35.
11. J. V. Neel, Diabetes mellitus: a 'thrifty' genotype rendered detrimental by 'progress'? *Am J Hum Genet* 1962/14: 353–62.
12. D. C. Benyshek and J. T. Watson, Exploring the thrifty genotype's food-shortage assumptions: a cross-cultural comparison of ethnographic accounts of food security among foraging and agricultural societies. *Am J Phys Anthropol* 2006/Epub ahead of print: 1–7.
13. Examples of how small changes (polymorphisms) in genes can affect susceptibility to metabolic disease include: J. Eriksson et al., The effects of the Pro12Ala polymorphism of the PPAR-γ2 gene on lipid metabolism interact with body size at birth. *Clin Genet* 2003/64: 366–70; E. Kajantie et al., The effects of the ACE gene insertion/deletion polymorphism on glucose tolerance and insulin secretion in elderly people are modified by birth weight. *J Clin Endocrinol Metab* 2004/89: 5738–41; A. Kubaszek et al., The association of the K121Q polymorphism of the plasma cell glycoprotein-1 gene with type 2 diabetes and hypertension depends on size at birth. *J Clin Endocrinol Metab* 2004/89: 2044–7; M. N. Weedon et al., Genetic regulation of birth weight and fasting glucose by a common polymorphism in the islet cell promoter of the glucokinase gene. *Diabetes* 2005/54: 576–81.
14. There have been many supporting epidemiological studies, generally involving surrogate endpoints such as blood pressure or insulin sensitivity (which itself is usually measured indirectly by testing how well the body metabolizes glucose; see n. 42 below). Most studies have confirmed the original disease-focused associations, but some confusion has arisen in the literature because there has been argument over the magnitude of the correlation found between these surrogates and birth size (R. Huxley et al., Unravelling the fetal origins hypothesis: is there really an inverse association between birthweight and subsequent blood pressure? *Lancet* 2002/360: 659–65). This is a very sterile argument; the issue of the magnitude of a statistical correlation between two surrogate markers (for birth weight is solely a surrogate for estimating the fetal environment) is irrelevant to the basic conclusion which is well proven clinically and experimentally—namely that changes in the fetal environment will affect the physiology of the offspring with long-term consequences. We propose that such long-term effects are driven by predictive responses and need not even involve changes in birth weight. The latter are driven by immediate adaptive needs. However, had it not been for the original birth weight observations, this whole field would not have been uncovered. For recent reviews of this subject see P. D. Gluckman and M. A. Hanson, Developmental origins of disease paradigm: a mechanistic and evolutionary

perspective. *Pediatr Res* 2004/56: 311–17, and P. D. Gluckman et al., Environmental influences during development and their later consequences for health and disease: implications for the interpretation of empirical studies. *Proc Royal Soc Lond B* 2005/272: 671–7.

15. P. D. Gluckman and M. A. Hanson (eds.), *Developmental Origins of Adult Disease*. Cambridge: Cambridge University Press, 2006.
16. K. A. Lillycrop et al., Dietary protein restriction of pregnant rats induces and folic acid supplementation prevents epigenetic modification of hepatic gene expression in the offspring. *J Nutr* 2005/135: 1382–6; I. C. G. Weaver et al., Epigenetic programming by maternal behavior. *Nat Neurosci* 2004/7: 847–54.
17. D. Chali et al., A case-control study on determinants of rickets. *Ethiop Med J* 1998/36: 227–34.
18. J. G. Eriksson et al., Pathways of infant and childhood growth that lead to type 2 diabetes. *Diabetes Care* 2003/26: 3006–10.
19. See P. D. Gluckman and M. A. Hanson, *The Fetal Matrix: Evolution, Development, and Disease*. Cambridge: Cambridge University Press, 2005.
20. The experimental skill of the German embryologist Hans Spemann (1869–1941) in manipulating amphibian eggs and embryos led him to discover that particular parts of the embryo, which he referred to as organizers, could direct the development of structures such as limbs even when transplanted to inappropriate areas. For this concept of 'embryonic induction' he was awarded the Nobel Prize in Physiology or Medicine for 1935. The American anatomist Charles R. Stockard (1879–1939), working at Cornell Medical College in New York, studied the effects of the external environment on the development of the embryos of a number of species. By changing the ionic composition of the growth medium or by treating the growing embryos with poisons such as alcohol, he was able to show that the effects of such factors depended on the timing of their application, thus defining 'critical moments' in development.
21. Worldwide, thalidomide caused about 12,000 babies to be born with deformities including lack of limbs. Dr William McBride's suspicions that thalidomide was the cause of these birth defects were initially received with scepticism by both the scientific community and the manufacturer of the drug. In 1961, *The Lancet* finally published a letter stating his claims and the drug was removed from the market. When the scale of the damage done by thalidomide became clear, McBride shot to fame and went on to receive honours and high recognition. In the early 1980s, McBride came to believe that another anti-nausea drug, Debendox, was also responsible for birth defects. To support this claim he published a paper showing that scopolamine, a chemical related to the active constituent in Debendox, caused deformities in offspring of rabbits. However, his assistants soon realized that some of the experiments he claimed to have performed had not actually taken place and in 1987 he was accused of fraud. In 1993 he was found guilty by a tribunal and was struck off the medical register (although

he was not banned from scientific research—there are no mechanisms to prevent a fraudulent scientist from working!). After initially denying the charges, McBride eventually conceded that he had falsified the results and was reinstated in 1998, when he returned to medical practice in Australia at the age of 71.

22. Growth hormone (GH), produced by the pituitary gland in the brain, is the primary determinant of growth in stature (linear growth) after birth. Its actions are in large part mediated by insulin-like growth factor (IGF)–1 produced either by the liver or locally in tissues. In contrast, fetal growth does not depend directly on GH; although the fetal pituitary gland does make the hormone, the necessary receptors in the fetal tissues only appear late in pregnancy. Rather, fetal growth appears to be driven directly by IGF–1 levels in the fetal and maternal circulations within the limitations set by nutrient supply across the placenta. Another related growth factor, IGF–2, is involved in growth of the embryo in the very early stages of pregnancy. GH does have a role in pregnancy, but in the mother—early in pregnancy, maternal GH production switches from the pituitary to the placenta (the placental GH is actually slightly different from pituitary GH) and its main role, together with another similar placental hormone called placental lactogen, appears to be to alter maternal insulin sensitivity, thus allowing the mother to utilize fatty acids as fuel and leaving glucose to be transferred across the placenta (P. D. Gluckman and C. S. Pinal, Regulation of fetal growth by the somatotrophic axis. *J Nutr* 2003/133: 1741S–1746S).
23. A. Walton and J. Hammond, The maternal effects on growth and conformation in Shire horse–Shetland pony crosses. *Proc Royal Soc Lond B* 1938/125: 311–35.
24. A. A. Brooks et al., Birth weight: nature or nurture? *Early Hum Dev* 1995/42: 29–35.
25. For example, otherwise well-nourished western women who eat a high-carbohydrate diet early in pregnancy and/or a low-protein diet late in pregnancy give birth to smaller babies (K. Godfrey et al., Maternal nutrition in early and late pregnancy in relation to placental and fetal growth. *BMJ* 1996/312: 410–14).
26. Z. A. Bhutta et al., Nutrition as a preventive strategy against adverse pregnancy outcomes. *J Nutr* 2003/133: 1589S–1767S.
27. J. F. Clapp III and E. L. Capeless, Neonatal morphometrics after endurance exercise during pregnancy. *Am J Obstet Gynecol* 1990/163: 1805–11.
28. M. Susser and Z. Stein, Timing in prenatal nutrition: a reprise of the Dutch Famine Study. *Nutr Rev* 1994/52: 84–94.
29. S. M. B. Morton, Maternal nutrition and fetal growth and development. In P. Glackman and M. Hanson (eds.) *Developmental Origins of Health and Disease* (Cambridge, Cambridge University Press, 2006).
30. Much of Ch. 3 discusses mechanisms by which intergenerational effects can arise through the mother, but can the father's history or environment affect his children? And if so, by what mechanism, since sperm is formed continually during a man's life? We know from clinical and experimental studies that mutations induced by exposure of males to chemicals or radiation can cause birth defects in

their offspring. More subtly, the large number of cell divisions involved in spermatogenesis suggests an opportunity for genetic or epigenetic damage to the precursor cells of sperm, and such damage might accumulate during ageing (R. L. Glaser and E. W. Jabs, Dear old dad. *Sci Aging Knowledge Environ* 21 Jan. 2004: 2004(3):re1). Indeed, there are some intriguing hints of a link between a man's age when he fathers a child and subsequent disease in that child—for example, the relationship between paternal age and schizophrenia (A. Sipos et al., Paternal age and schizophrenia: a population based cohort study. *BMJ* 2004/329: 1070), although in such studies it is always difficult to disentangle inherited from environmental factors. There is epidemiological evidence for intergenerational effects transmitted through the male line—linking childhood smoking by fathers with obesity in their sons, and nutrition of grandfathers with mortality in their grandsons—with exposure in the pre-pubertal slow growth period being critical (M. E. Pembrey et al., Sex-specific, male-line transgenerational responses in humans. *Eur J Hum Genet* 2006/14: 159–66). But our understanding of the influence of the father's environment was recently advanced by a study of the transgenerational effects of endocrine disruptors (see Ch. 6 n. 27 and Ch. 8 n. 21) on male fertility in rats. Pregnancy in these animals lasts twenty-one or twenty-two days, and sex determination and testis development in their fetuses occur over days 12–15. If pregnant rats are treated with endocrine disruptors during this period, but not if they are treated later in pregnancy, their male offspring show defects in sperm formation and fertility which are transferred through the male line to at least the fourth generation. These defects correlate with epigenetic (DNA methylation) alterations in the testis. Thus, transient exposure to an environmental toxin at a critical stage in development causes a disease state that can be transmitted to subsequent generations through the paternal line (M. D. Anway et al., Epigenetic transgenerational actions of endocrine disruptors and male fertility. *Science* 2005/308: 1466–9). The implications of this study for human health are considerable.

31. P. D. Gluckman and M. A. Hanson, Living with the past: evolution, development and patterns of disease. *Science* 2004/305: 1773–6; P. D. Gluckman et al., Environmental influences during development and their later consequences for health and disease: implications for the interpretation of empirical studies. *Proc Royal Soc Lond B* 2005/272: 671–7.
32. For example we sweat when we have a fever to try to reduce our temperature, we shiver when we are cold to generate heat. The fetus does neither because its temperature is controlled by the temperature of the uterus which is in turn set by maternal state. Only at birth must the fetus start controlling its own temperature. Animals including humans have the capacity to start generating heat very soon after birth, when chemical signals from the placenta such as adenosine which inhibits heat production are removed by tying the umbilical cord. But thermal control in newborn babies can be poor—that is why the invention of the

incubator at the beginning of the 20th century was a critical step in the development of newborn care. Newborn lambs are only born with enough fuel to burn for heat for about twelve to twenty-four hours. That is why they will die of hypothermia not only under extreme thermal conditions but also if they cannot suckle well enough to get sufficient nutrition to sustain heat production.

33. P. Bateson, Fetal experience and good adult design. *Int J Epidemiol* 2001/30: 928–34; P. D. Gluckman and M. A. Hanson, Living with the past: evolution, development and patterns of disease. *Science* 2004/305: 1773–6; P. D. Gluckman et al., Environmental influences during development and their later consequences for health and disease: implications for the interpretation of empirical studies. *Proc Royal Soc Lond B* 2005/272: 671–7; P. D. Gluckman et al., Predictive adaptive responses and human evolution. *Trends Ecol Evol* 2005/20: 527–33.
34. For example, see Introduction n. 17.
35. This long-lived change in behaviour involves epigenetic mechanisms—specifically, DNA methylation of the molecular switch (promoter) controlling expression of a stress hormone receptor in the brain. Moreover, the behaviour and the DNA methylation can be restored in grown-up rats by treatment with high levels of the essential amino acid necessary for DNA methylation (I. C. Weaver et al., Reversal of maternal programming of stress responses in adult offspring through methyl supplementation: altering epigenetic marking later in life. *J Neurosci* 2005/25: 11045–54).
36. K. A. Lillycrop et al., Dietary protein restriction of pregnant rats induces and folic acid supplementation prevents epigenetic modification of hepatic gene expression in the offspring. *J Nutr* 2005/135: 1382–6.
37. P. D. Gluckman and M. A. Hanson, *The Fetal Matrix: Evolution, Development, and Disease*. Cambridge: Cambridge University Press, 2005.
38. See Ch. 3 n. 26.
39. S. M. Robinson et al., Impact of educational attainment on the quality of young women's diets. *Eur J Clin Nutr* 2004/58: 1174–80.
40. H. Takimoto et al., Increase in low-birth-weight infants in Japan and associated risk factors, 1980–2000. *J Obstet Gynaecol Res* 2005/31: 314–22; personal communication from Dr K. Ueda.
41. For an extensive discussion of this theory see P. D. Gluckman and M. A. Hanson, *The Fetal Matrix: Evolution, Development, and Disease*. Cambridge: Cambridge University Press, 2005.
42. The body normally copes with an increase in blood glucose after a meal by secreting insulin from the pancreas. In turn, insulin stimulates the transport of glucose into muscle and other tissues. Damping down this response by reducing the sensitivity of muscle to insulin, thereby maintaining a higher blood glucose level, is advantageous if food is limited—but if food is not limited, the persistently high blood glucose levels are damaging and indeed this is the pattern seen in type 2 diabetes. So decreased insulin sensitivity (also called insulin resistance) is

considered a prodrome for the later development of type 2 diabetes. A recent study has shown that babies born small actually have increased insulin sensitivity at birth and then develop insulin resistance over the first three years of life—which suggests that this metabolic adjustment really is an adaptive response to the anticipated environment rather than just a short-term solution to adverse prenatal conditions (V. Mericq et al., Longitudinal changes in insulin sensitivity and secretion from birth to age three years in small- and appropriate-for-gestational-age children. *Diabetologia* 2005/48: 2609–14).

43. For a more complete discussion of the types of response the fetus will make to predicting a poor postnatal environment, see P. D. Gluckman and M. A. Hanson, *The Fetal Matrix: Evolution, Development, and Disease*. Cambridge: Cambridge University Press, 2005.
44. The main source of data about prolonged infant undernutrition comes from the studies of Eriksson and colleagues in Finland. Their studies suggest that heart disease and diabetes are more common in those children born smaller, who stay small and do not show catch-up growth until the third year of life. Subsequently these children put on weight faster than those who do not later develop heart disease (D. J. Barker et al., Trajectories of growth among children who have coronary events as adults. *N Engl J Med* 2005/353: 1848–50). But Alan Lucas and his colleagues have presented several studies in which infants fed a cow's milk-based bottle formula feed (which has much higher energy and protein levels than breast milk) put on weight quickly in infancy and later have biochemical markers that suggest that they are at greater risk of obesity and metabolic disease in later life (A. Singhal et al., Early nutrition and leptin concentrations in later life. *Am J Clin Nutr* 2002/75: 993–9; A. Singhal et al., Low nutrient intake and early growth for later insulin resistance in adolescents born preterm. *Lancet* 2003/361: 1089–97; A. Singhal et al., Breastmilk feeding and lipoprotein profile in adolescents born preterm: follow-up of a prospective randomised study. *Lancet* 2004/363: 1571–8). Recent systematic reviews of all the available studies show that formula feeding is associated with a greater risk of later obesity than is breast feeding (C. G. Owen et al., Effect of infant feeding on the risk of obesity across the life course: a quantitative review of published evidence. *Pediatrics* 2005/115: 1367–77; T. Harder et al., Duration of breastfeeding and risk of overweight: a meta-analysis. *Am J Epidemiol* 2005/162: 397–403).
45. The International Diabetes Federation has recently developed new criteria for the diagnosis of metabolic syndrome (K. G. M. M. Alberti et al., The metabolic syndrome: a new worldwide definition. *Lancet* 2005/366: 1059–62). Central obesity, as assessed by waist circumference, is essential to the diagnosis, accompanied by any two of high blood triglyceride level, reduced high density lipoprotein cholesterol ('good' cholesterol), raised blood pressure, and raised fasting plasma glucose (indicating insulin resistance). Using this definition, the prevalence of metabolic syndrome in adults varies from 8% in Hong Kong

Chinese, through 20–30% in various European populations, to 39% in the USA.

46. N. Stettler et al., Early risk factors for increased adiposity: a cohort study of African American subjects followed from birth to young adulthood. *Am J Clin Nutr* 2000/72: 378–3; K. K. Ong et al., Size at birth and early childhood growth in relation to maternal smoking, parity and infant breast-feeding: longitudinal birth cohort study and analysis. *Pediatr Res* 2002/52: 863–7.
47. R. C. Whitaker, Predicting preschooler obesity at birth: the role of maternal obesity in early pregnancy. *Pediatrics* 2004/114: 29–36; K. K. Ong and D. B. Dunger, Birth weight, infant growth and insulin resistance. *Eur J Endocrinol* 2004/151: U131–U139.
48. J. G. Eriksson et al., Effects of size at birth and childhood growth on the insulin resistance syndrome in elderly individuals. *Diabetologia* 2002/45: 342–8.
49. K. Srinath-Reddy et al., Responding to the threat of chronic diseases in India. *Lancet* 2005/366: 1744–9.
50. C. H. Fall et al., Micronutrients and fetal growth. *J Nutr* 2003/133 (5 Suppl 2): 1747S–1756S.
51. The 'small but healthy' hypothesis, originally associated with David Seckler, proposed that since a small body size requires less nutrients, stunted growth is an adaptation that permits individuals and populations to function at levels of nutrition that are too low to support larger body size (D. Seckler, 'Small but healthy': a basic hypothesis in the theory, measurement and policy of malnutrition. In P. V. Sukhatme (ed.), *Newer Concepts in Nutrition and their Implications for Policy*. Pune: Maharashtra Association for the Cultivation of Science, 1982; D. Seckler, Malnutrition: an intellectual odyssey. In K. T. Achaya (ed.), *Interfaces between Agriculture, Nutrition and Food Science*. Tokyo: United Nations University, 1984).
52. J. C. Waterlow, Nutritional adaptation in man: general introduction and concepts. *Am J Clin Nutr* 1990/51: 259–63. The economist Sir Partha Dasgupta discusses the idea in a broader context in his account of the economics of poverty *An Inquiry into Well-Being and Destitution* (Oxford: Clarendon Press, 1993).
53. See n. 52 above.
54. P. D. Gluckman et al., Life-long echoes: a critical analysis of the developmental origins of adult disease model. *Biol Neonate* 2004/87: 127–39.
55. World Health Organization, Promoting optional fetal development: report of a technical consultation. (Geneva, World Health Organization, 2006).
56. C. Power and B. J. Jefferis, Fetal environment and subsequent obesity: a study of maternal smoking. *Int J Epidemiol* 2002/31: 413–19.
57. M. H. Vickers et al., Neonatal leptin treatment reverses developmental programming. *Endocrinology* 2005/146: 4211–16.
58. For example, a polymorphism in the PPAR-γ2 gene, which codes for a nuclear receptor which regulates the growth and metabolism of adipose tissue, only leads

to a greater risk of diabetes in those born small (J. Eriksson et al., The effects of the Pro121A polymorphism of the PPAR-γ2 gene on lipid metabolism interact with body size at birth. *Clin Genet* 2003/64: 366–70).

59. We will leave discussion of the one possible exception to this statement until Ch. 8.

Chapter 8

1. This calculation is fraught with difficulties (see Ch. 4 n. 35) but there is a remarkable consensus: see S. N. Austad, Menopause: an evolutionary perspective. *Exp Gerontol* 1994/29: 255–63; R. H. Steckel and J. C. Rose (eds.), *The Backbone of History: Health and Nutrition in the Western Hemisphere*. Cambridge: Cambridge University Press, 2002.
2. R. W. Fogel, *The Escape from Hunger and Premature Death, 1700–2100: Europe, America and the Third World*. Cambridge: Cambridge University Press, 2004.
3. M. R. Rose, *Evolutionary Biology of Aging*. New York: Oxford University Press, 1991.
4. See Ch. 3 n. 20.
5. W. G. Sumner, *Folkways*. New York: Ginn & Co., 1907.
6. K. McComb et al., Matriarchs as repositories of social knowledge in African elephants. *Science* 2001/292: 491–4.
7. J. Kingdon, *Lowly Origins: Where, When and Why our Ancestors First Stood Up*. Princeton: Princeton University Press, 2003.
8. T. T. Perls and R. C. Fretts, The evolution of menopause and human life span. *Ann Hum Biol* 2001/28: 237–45.
9. J. Vijg and Y. Suh, Genetics of longevity and aging. *Annu Rev Med* 2005/56: 193–212.
10. S. E. Ozanne and C. N. Hales, Lifespan: catch-up growth and obesity in male mice. *Nature* 2004/427: 411–12.
11. H. Van Remmen et al., The anti-ageing action of dietary restriction. *Novartis Found Symp* 2001/235: 221–30. This difference between prenatal and postnatal nutritional effects on longevity highlights how critical it is to interpret environmental cues in relation to the developmental stage at which the cues act—this important point is not always well understood by experimentalists.
12. R. H. Post, Early menarchial age of diabetic women. *Diabetes* 1962/11: 287–90.
13. E. Kajantie et al. Size at birth as a predictor of mortality in adulthood: a follow-up of 350 000 person-years. *Int J Epidemiol* 2005/34: 655–63.
14. There are some important genetic interactions in this relationship. With one particular polymorphism of the vitamin D receptor gene, the relationship is reversed (C. Cooper et al., Review: developmental origins of osteoporotic fracture. *Osteoporos Int* 2006/17: 337–47).

15. F. Nottebohm, The road we travelled: discovery, choreography, and significance of brain replaceable neurons. *Ann NY Acad Sci* 2004/1016: 628–58.
16. C. Tolsa et al., Early alteration of structural and functional brain development in premature infants born with intrauterine growth restriction. *Pediatr Res* 2004/56: 132–8.
17. C. M. McCarton, et al. Cognitive and neurologic development of the premature, small for gestational age infant through age 6: comparison by birth weight and gestational age. *Pediatrics* 1996/98: 1167–78; I. Elgen et al., A non-handicapped cohort of low-birthweight children: growth and general health status at 11 years of age. *Acta Paediatr* 2005/94: 1203–7.
18. T. L. Spires and A. J. Hannan, Nature, nurture and neurology: gene–environment interactions in neurodegenerative disease. *FEBS J* 2005/272: 2347–61.
19. M. Greaves, Cancer causation: the Darwinian downside of past success? *Lancet Oncol* 2002/3: 244–51.
20. T. Flannery, *Country*. Sydney: Text Publishing, 2004.
21. Endocrine disruptors are environmental chemicals that produce unwanted physiological effects by their influence on hormonal regulation, either by binding directly to hormone receptors or by altering the synthesis or metabolism of endogenous hormones. For example, many chemicals, including some natural products (such as genistein in soy) or some industrial chemicals (for example, bisphenol A and the phthalate plasticisers), can be shown to bind to oestrogen receptors and elicit the effects of the natural oestrogens. There are data that suggest that inappropriate exposure of males to such xenobiotics and toxins might cause feminization of the fetus and possibly infertility in the adult, whereas increased exposure of females to oestrogens may have a role in the development of breast cancer (R. H. Waring and R. M. Harris, Endocrine disrupters: a human risk? *Mol Cell Endocrinol* 2005/244: 2–9). The role and importance of endocrine disruptors in human disease remains highly controversial (S. Safe, Clinical correlates of environmental endocrine disruptors. *Trends Endocrinol Metab* 2005/16: 139–44). See also Ch. 6 n. 27 and Ch. 7 n. 30.
22. M. Greaves, Cancer causation: the Darwinian downside of past success? *Lancet Oncol* 2002/3: 244–51.
23. R. Kaaks, Nutrition, insulin, IGF-1 metabolism and cancer risk: a summary of epidemiological evidence. *Novartis Found Symp* 2004/262: 247–60.
24. S. I. Dos Santos et al., Is the association of birth weight with premenopausal breast cancer risk mediated through childhood growth? *Br J Cancer* 2004/91: 519–24.
25. A. D. L. Napier, *The Menopause and its Disorders*. London: Scientific Press, 1987.
26. N. Howell, *Demography of the Dobe !Kung*. New York: Academic Press, 1979.
27. H. Marsh and T. Kasuya, Changes in the ovaries of the short-finned pilot whale, *Globicephala macrorhynchus*, with age and reproductive activity. *Rep Int Whale Comm Special* 1984/6: 311–35.

28. S. N. Austad, Menopause: an evolutionary perspective. *Exp Gerontol* 1994/29: 255–63.
29. As shown by a study of gorillas in North American zoos: **http://news.nationalgeographic.com/news/2005/12/1227_051227_gorilla.html**.
30. G. C. Williams, Pleiotropy, natural selection, and the evolution of senescence. *Evolution* 1957/11: 398–411.
31. K. Hawkes et al., Grandmothering, menopause, and the evolution of human life histories. *Proc Natl Acad Sci USA* 1998/95: 1336–9.
32. R. Sear et al. Maternal grandmothers improve nutritional status and survival of children in rural Gambia. *Proc R Soc Lond B* 2000/267: 1641–7.
33. M. Lahdenpera et al., Fitness benefits of prolonged post-reproductive lifespan in women. *Nature* 2004/428: 178–81.
34. J. Kingdon, *Lowly Origins: Where, When and Why our Ancestors First Stood Up*. Princeton: Princeton University Press, 2003; R. H. Steckel and J. C. Rose (eds.), *The Backbone of History: Health and Nutrition in the Western Hemisphere*. Cambridge: Cambridge University Press, 2002.
35. D. P. Shanley and T. B. L. Kirkwood, Evolution of the human menopause. *Bioessays* 2001/23: 282–7.

Chapter 9

1. A. Prentice, Constituents of human milk. In *Breastfeeding: Science and Society*. Tokyo: United Nations University, 1996. Available online from: **www.unu.edu/unupress/food/8F174e/8F174E04.htm_Constituents%20of%20human%20milk**.
2. R. Morley, et al., Neurodevelopment in children born small for gestational age: a randomized trial of nutrient-enriched versus standard formula and comparison with a reference breastfed group. *Pediatrics* 2004/113: 515–21; C. J. Chantry et al., Full breastfeeding duration and associated decrease in respiratory tract infection in US children. *Pediatrics* 2006/117: 425–32. See also Ch. 7 n. 44, for a discussion of the metabolic consequences of bottle-feeding.
3. F. A. Young et al., The transmission of refractive errors within Eskimo families. *Am J Optom Arch Am Acad Optom* 1969/46: 676–85.
4. The mechanism of regulating eye growth in relation to focus is termed emmetropization. There are a number of genetic causes of severe (high) myopia but these are rare. While it was previously thought that juvenile onset myopia had a major genetic component, the more recent consensus is that environmental factors are much more important (I. Morgan and I. Rose, How genetic is school myopia? *Progr Retinal Eye Res* 2005/24: 1–38).
5. L. L. Lin et al., Prevalence of myopia in Taiwanese schoolchildren: 1983 to 2000. *Ann Acad Med Singapore* 2004/33: 27–33.
6. Strictly speaking, the studies are not clear on the actual environmental factor

leading to the greater incidence of myopia. Reading seems the most obvious factor, but artificial lighting and even dietary changes may also be involved (see reference in n. 4 above).

7. See Ch. 4 n. 51.
8. S. Baron-Cohen (ed.), *The Maladapted Mind: Classic Readings in Evolutionary Psychopathology*. Hove: Psychology Press, 1997.
9. R. A. Foley, The adaptive legacy of human evolution: a search for the environment of evolutionary adaptedness. *Evol Anthropol* 1996/4: 194–203.
10. Whilst speculative it is interesting to consider whether the fundamental processes of developmental plasticity might themselves be subject to environmental influences. Could abnormal environmental factors shift the trajectory of development to create an altered risk of mismatch? We give this question some emphasis because of the growing evidence that some toxins such as bisphenol A found in plastics might induce epigenetic change. In other words, even chemicals that are at sub-toxic levels in the environment might, if they produced small changes in the setting of the normal developmental adaptive responses, produce a major shift in the number of people who fall into the 'mismatched' as opposed to the 'matched' category, and thus have a dramatic effect on the number of people in the population who are unhealthy. Similar considerations apply to many other factors. For example, there is evidence that children born to mothers who smoke are at greater risk of developing obesity (C. Power and B. J. Jefferis, Fetal environment and subsequent obesity: a study of maternal smoking. *Int J Epidemiol* 2002/31: 413–19). Is this simply due to smoking causing the fetus to predict a nutritionally poor postnatal environment (because nicotine interferes with nutritional transport from mother to fetus) or is there a toxic component as well? The study of such environmental interactions during development is a new but important item on the research agenda.
11. M. Horan et al., Human growth hormone 1 (GH1) gene expression: complex haplotype-dependent influence of polymorphic variation in the proximal promoter and locus control region. *Hum Mutat* 2003/21: 408–23.
12. M. Crump, *In Search of the Golden Frog*. Chicago: University of Chicago Press, 2000; T. Flannery, *The Weather Makers: The History and Future Impact of Climate Change*. Sydney: Text Publishing, 2005. Recent research suggests another way in which climate change may lead to frog extinction: the changed climatic conditions in Monteverde favour the growth of a fungus which dehydrates their amphibian victims—this fungus is thought to have been involved in the extinction of many harlequin frog species (J. A. Pounds et al., Widespread amphibian extinctions from epidemic disease driven by global warming. *Nature* 2006/439: 161–7).
13. Robert Koch (1843–1910) worked initially with anthrax to show that bacilli caused the infection—his early studies were carried out in his apartment because he did not have a laboratory. He later discovered the organisms that cause tuberculosis and cholera and was awarded the Nobel Prize in 1905.

14. See Ch. 8 n. 14.
15. See Introduction n. 14 and Ch. 1 n. 15 for descriptions of how the concepts discussed in these two paragraphs developed.
16. R. M. Nesse and G. C. Williams, *Why We Get Sick: The New Science of Darwinian Medicine*. New York: Vintage, 1994. Other books on evolutionary medicine include: W. R. Trevathan et al. (eds.), *Evolutionary Medicine*. New York: Oxford University Press, 1999; S. C. Stearns (ed.), *Evolution in Health and Disease*. Oxford: Oxford University Press, 1999.
17. See Ch. 7 n. 57.
18. Such effects can be observed in both animals (D. Liu et al., Maternal care, hippocampal synaptogenesis and cognitive development in rats. *Nat Neurosci* 2000/3: 799–806) and humans (R. Feldman et al., Comparison of skin-to-skin (kangaroo) and traditional care: parenting outcomes and preterm infant development. *Pediatrics* 2002/110: 16–26).
19. The Indian medical research community has shown a great interest in the issues of developmental and metabolic mismatch. The oldest scientific society devoted to the study of the matters discussed in Chapter 8 is the Society for the Study of the Natal Effects on Health of Adults, SNEHA, an Indian medical group which has had regular meetings for well over a decade. Their meetings show a breadth of knowledge and approach and a desire to marry science with action that many other academic societies could learn much from.
20. See Introduction n. 8.

Acknowledgements

To address the questions we raise in this book we had to explore different scientific worlds, ranging from molecular genetics and evolutionary biology to social anthropology and public health. As well as our own research and that of our research teams we have benefited from many interactions with colleagues in several countries. In evolutionary biology we particularly want to thank David Raubenheimer (Auckland), Hamish Spencer (Otago), David Haig (Harvard), and Eva Jablonka (Tel Aviv). In developmental biology we are indebted to Scott Gilbert (Swarthmore), Brian Hall (Halifax), and Patricia Monaghan (Glasgow). Within the anthropological community we particularly thank Marta Lahr and Rob Foley (Cambridge), Barry Bogin (Michigan), Chris Kuzawa (Northwestern), and Peter Ellison (Harvard). In reproductive and fetal biology we want to thank Mark Vickers, Jane Harding, Frank Bloomfield, Bernie Breier, Mark Oliver, Anne Jacquiry, Tom Fleming, Lucy Green, Felino Cagampang, Rohan Lewis, Judith Eckert, Kirsten Poore, Chris Torrens, Caroline Bertram, Jane Cleal, Lucy Braddick, Paul Terroni, Adam Watkins, Jo Rodford, Julian Boullin, Omar Khan, Tristram Snelling, Ryan Chau, Junlong Zhang, and Simon Cunningham in our own universities. In epigenetics and nutrition we thank Alan Jackson, Karen Lillycrop, Graham Burdge, and Fred Anthony (Southampton), Terrence Forrester (Kingston, Jamaica), Wolf Reik (Cambridge), Jonathon Seckl (Edinburgh), Michael Meaney (Montreal), and Allan Sheppard (Hamilton, New Zealand). In physiology and endocrinology we thank John Challis and Steve Matthews (Toronto), Geraldine Clough (Southampton), Dino Giussani (Cambridge), Anibal Llanas (Chile), Ray Noble, Paul Taylor, and Lucilla Poston (London). In clinical medicine we thank Mel Grumbach (San Francisco), John Newnham (Perth), Wayne Cutfield and Paul Hofman (Auckland), Carlos Blanco (Maastricht), Michael Ross (San Francisco), Guttorm Haugen (Oslo), Torvid Kiserud (Bergen), Christopher Byrne, Iain Cameron, Sunil Ohri, Tim Wheeler, and Graham Roberts (Southampton). Tom Kirkwood (Newcastle-upon-Tyne) was generous in his advice on ageing and Stan Neuman (UCL) commented on psychological perspectives. We thank Harold Alderman (World Bank) for his economic insights and Zulfi Bhutta (Karachi) for his focus on the issues of public health priorities. We thank our many colleagues and friends in India and in particular Ranjan Yajnik (Pune), Harshi Sachdev, and Santosh Bhargava (Delhi) for their remarkable insights into the impact of the nutritional transition. Amongst the epidemiological community we particularly want to thank our colleagues Susan Morton (Auckland), Johan Eriksson (Helsinki), Catherine Law (London) Keith Godfrey, Cyrus Cooper, Caroline Fall, Hazel Inskip, Clive Osmond, and

David Phillips (Southampton), Hiroaki Fukuoka (Tokyo), and particularly David Barker (Southampton).

Sir Patrick Bateson FRS, formerly provost of King's College, Cambridge, and Vice President and Biological Secretary of the oldest of the great scientific academies, the Royal Society, played no small part in starting us on this journey into a new field of evolutionary biology and medical science, and his continued mentorship and guidance and his advice on the manuscript have been much appreciated.

We are particularly grateful for Robert Winston's continuing interest in our work and his willingness to write a foreword.

Deborah Burrage and Alan Beedle did much of the fact checking and helped enormously with preparing the notes. Alan put much effort into the final editing. Hamish Spencer, Andrea Graves, Pat Bateson, Judy Gluckman, and David Raubenheimer devoted considerable time to reading the manuscript and suggesting improvements. Deborah Peach repeatedly decoded our handwriting and processed versions of the book during its gestation. Again we thank our universities for supporting this joint endeavour. We gratefully acknowledge the support of the British Heart Foundation and the Hood Fellowship Fund of the University of Auckland. We thank our agent Mandy Little for her unfailing support, and our editor Latha Menon and production assistant James Thompson at OUP for their assistance and encouragement.

The journey that started this book began as two separate journeys, to two extreme places. One of us was breathless and cursing as he struggled up a mountainside in the Himalayas, the other feverish and dehydrated in the semi-desert of sub-Saharan West Africa. Both were undertaking medical research for the first time. That was over thirty years ago, but it has led us to the recognition that the nature of our life's journey, and its consequences for each and every one of us, very much depends on avoiding *mismatch*.

P.D.G
M.A.H.

Auckland, Southampton, and places in between
February 2006

Index

adaptation:
  camel's hump as 160
  definition 42
  stunting 175
  to iodine deficiency in Sherpa 43
  to occupy niches 45, 97
adaptationist:
  analysis of human disease 203
  criticism of approach 19
  extension into sociobiology 198
adaptedness 6
adaptive range 126, 201, 202
adaptive responses:
  immediate 82–3, 167
  in human biology 201
  life-history strategy 130
  long-term 130
  predictive, *see* predictive adaptive responses
  short-term 130–1
adolescence:
  as transition from child to adult 138
  behaviour 150, 151
  dependency 139
  independence at end of 141
  puberty rituals 138
  rapid growth during 84
adult-onset diabetes *see* diabetes
Africa:
  agriculture 104
  bilharzia 95
  butterfly 87
  chimpanzees 70, 141
  elephants 86, 103, 188
  human evolution 6, 29, 84–5, 94, 100, 103
  human migration from 43, 47, 50, 100, 122
  hunter-gatherers 111, 188
  tool-making by hominids 99
African-Americans 127
ageing:
  allocation of health care resources 193
  as mismatch 186
  as trade-off 184–5
  degenerative diseases 186
  human 44
  mechanisms 86, 118, 181–3
  of human eggs 76, 190
  stem cells 77
agouti mouse 67–8
agriculture 73, 102, 104–5, 109
Alaska 161
Albigensian Crusade 158
alleles 20
altricial 92
altruism 28
Alzheimer's disease 186
America *see* North America, South America, USA
art 101
artificial selection 23, 25, 59, 104
assisted reproductive technology 117, 165
asthma 106
Australia:
  Aborigines, *see* Australian Aborigines
  agriculture 104, 183
  child health 113
  colonization 100
  degradation of environment 8
  koala 39
  marsupial mouse 140
  monotremes 93
  poison pea 187
  rabbits 38–9
  redback spider 139
  settlement by humans 101

skin cancer 187
termites 96
tiger snakes 33–4
Australian Aborigines:
age of weaning 133
arrival in Australia 100
effect of colonization 111
social structure 137, 183
use of technology 102
auxology 152
avian influenza 106, 114, 127

baboons:
ageing 182
learning 70
mating strategies 27
social structure 27, 57
survival to adulthood 142
'back to sleep' 73
Barker, David 163, 168
barnacles 26, 81
bar-tailed godwit 161
beak Galapagos finches 97
Belyaev, Dmitry Konstantinovich 25
Bhutta, Zulfiqhar 210
bilharzia 95–6
bipedality 92, 132
birds:
migration 161
renewal of brain cells 185
birth:
caesarian section 92
maternal age at first 116
size of pelvic canal 92
birth defects 79, 164
birth weight:
control by maternal constraint 132
Dutch Hunger Winter 68–9
effect on uterine size 69
Boas, Franz 113
Boroola mutation 19
bottle feeding:
developing countries 196
later consequences 172, 196
Bowlby, John 103
brain:
ageing 185
development 85, 129
human, immaturity at birth 92–3
pituitary gland 3
size at birth 140
timing of maturation 150
brain function:
gender differences 144
modular 103
breast cancer 188
breast feeding:
effect on birth spacing 149
effect on later obesity 172
human dependency 84
nutrition quality during extended 133
practices in different cultures 73
brown fat 9
butterfly 87

Caenorhabditis elegans 78
caesarian section 92
California 8, 57, 79, 81, 184
Canada 113, 191
cancer:
as consequence of ageing 186
as mismatch 186
effect of diet 187
carbohydrates 119
castrati 145
Cathars 158
childhood development 85
childhood obesity:
effect of birth order 173
India 120
recent appearance 120
reduction of physical activity 161
chimpanzees:
birth 92, 132
brain size 140
DNA sequence 30
evolution 29
learning 70
mating strategy 27
plant use for medication 114

social structure 57, 141
tool use 70, 141
China:
life expectancy 178
restriction of family size 116
vehicle ownership and obesity 161
chromosomes:
arrangement of genes on 22, 60, 61
copying 186
sex determination 20
X, *see* X chromosome
Y, *see* Y chromosome
cinchona tree 114
circumcision 137
cities 107–10
clan size 143
climate change 44, 99
cloaca 93
cognitive function:
impairment by undernutrition 175
optimization by maternal-infant interaction 207
cognitive impairment:
during ageing 186
in children born small 207
coinage 107
colonization 106, 110
comfort zone 33, 42–4, 82
communicable disease 114–15
conception:
age of genetic material at 76
as start of life 76
environmental signals at time of 68, 80, 198
nutritional status at time of 68, 80, 166, 170, 176, 206
constraint:
evolutionary 125–8
imposed by fidelity of predictive response 131
in design 128–32
maternal, *see* maternal constraint
of birth size by pelvic dimensions 85, 92, 132, 164
of genetic variation 24
consumer culture 179
consumer durables 180
contraception 116, 156
Costa Rica 200
costs:
of adaptation 12, 41–4, 175
of chosen life-history strategy 44
of developmental plasticity 81–2, 131
of healthcare 174, 176, 189, 192, 207, 209
see also trade-offs
cot death 73
cow's milk *see* bottle feeding
cretinism 3–4, 43, 200, 212
cultural evolution 69–73, 117, 156
cultural explosion 72
cultural inheritance 69–73, 207
cystic fibrosis 206

dairying 105
Daphnia 10
Darwin, Charles:
artificial selection 23, 59
barnacles 26
development of evolutionary theory 21–6
embryology 9
finches 97–8
influence by gradualism 46
natural selection 23, 64
The Descent of Man and Selection in Relation to Sex 55, 57
The Origin of Species 10, 46
The Voyage of the Beagle 8
variation 22
Darwinian evolution:
as possible solution to mismatch 208–9
fundamental tenets 53
occurrence in modern humans 126
Darwinian fitness:
definition 17, 26
life-history strategy 28
modern separation from reproduction 117

Darwinian medicine *see* evolutionary medicine
deer 55, 56, 88
degenerative disease 181
demography 192
design 7
development 9, 79–81, 198
developmental mismatch 130, 133, 202
developmental plasticity 35–8
  as source of variation between individuals 199
  definition 21, 30
  evolution of 168
  irreversibility 128
  limitations 130, 180
  matching individual to environment 79–81, 169, 198
  mediation by epigenetics 129
  role in determining constitution of the individual 200
  sensors for 128
  types of responses 130
diabetes:
  adult onset (type 2) 50, 159, 174
  as part of metabolic syndrome 173, 174
  consequence of low birthweight 163
  consequences of 173
  Ethiopian Jews 51
  fetal heart defects 79
  genetic factors influencing 50, 202
  increasing incidence 167, 176
  maternal 165, 166, 174
  maternal obesity as contributing factor 80
  metabolic mismatch 173, 195, 200
  nutritional transition 51, 119, 159, 207
  relation to age of puberty 184
  thrifty genotype 162
Dickens, Charles 110
diet:
  and cancer 187
  at conception 68, 170
  balanced as part of health promotion 205, 208
  change after development of agriculture 103, 105, 119
  during pregnancy 30, 68, 90, 131, 165, 166, 170, 171, 206
  Easter Island 8
  effect of agriculture on human 105
  experimental reversal of effects of poor 176, 206
  hunter-gatherer 107, 119
  improvement 205
  in city life 108, 110
  iodine deficiency 2, 3, 43, 212
  lactose content 105
  malocclusion 35
  maternal, effect on fetal calcium 30
  nutritional transition 119, 174, 204
  Palaeolithic 103–4, 118
  parental behaviour as cause of poor 51
  relevance to obesity 50, 51, 161, 173
  socioeconomic effects on quality 173
  treatment of maple syrup urine disease 30
  vitamin C deficiency and scurvy 12
discount rate 209
disease:
  as interaction between normal biology and normal environment 202
  distal causes 210
  infectious, after human settlement 105
  proximate causes 210
DNA methylation 62, 66–8, 206
DNA repair 186
DNA sequence 20–2
dodo 47
domestication 104
dominant inheritance 59
Dominican Republic 58
Dutch Hunger Winter 68–9

Easter Island 8
ecological developmental biology 10, 163, 203
economic modelling 209

INDEX

economic transition 159, 178, 204, 207
economics 122
eggs 76–7
elephants:
  African 86, 103, 188
  dwarf 64
  social role of matriarch 183
embryology 9, 203
embryonic development 164
embryonic diapause 88
embryonic stem cells 77
endocrine disruptors 148, 187
England:
  life expectancy 178
  social reform 153
  tuberculosis risk 115
  see also UK
Enlightenment 111–12, 153
environment:
  alteration by human activity 98–123, 126–8, 195
  and evolution 63–6
  and risk of disease 114
  during development 162
  nutritional, *see* nutritional environment
  postnatal 168
  range 8
environment of evolutionary adaptedness 103
environmental niche:
  adaptation to 6
  construction 98
  Galapagos finches 97–8
  giant panda 39
  in relation to neck length 54
  in relation to specialism 39, 41, 45, 47
  mountain vole 88
environmental signals:
  causing developmental plasticity 80, 88
  causing epigenetic change 31
  during development 37
  fetal response to 130, 163, 167–8, 177
  sensitivity of response to 177
environmental toxins 187
epidemiology 115, 163, 187, 201, 210
epigenetic inheritance 15, 51, 52, 67, 125, 131, 198, 206
epigenetic marks 62, 66–8, 206
epigenetic revolution 51, 66, 204
epigenetics:
  and cancer 66
  and nutrition 206
  definition 66
  effect on gene expression 35, 62, 67, 68, 78, 163, 166, 169, 203, 206
  in development 66, 166, 199, 204, 206
  modification of DNA structure 34, 62, 63, 67, 129, 169, 204, 206, 210
Ethiopia 163
eugenics 61
Europe:
  Neanderthals 45, 47
  peoples as colonizers 71, 102, 104, 110
  persistence of intestinal lactase 105
  population 109
  settlement by humans 43, 47
  skin cancer 44
evolution:
  cultural 69–73, 117, 156
  human 62, 70, 85, 102, 119, 127, 132, 146, 162, 171
  interaction with development 7, 9, 11, 32, 79, 124, 128, 202, 203
  mechanisms 17–19, 52–7
  of ageing 177, 179, 186
  of language 101
  of menopause 154, 188–91
  of religion 72, 114
  timescale 7, 42, 46, 208
evolutionary developmental biology 202
evolutionary medicine 203
exercise:
  as public health intervention 175, 205, 208
  children 174
  developing world 161
  economic transition 207

energy use 118
habit in family 50, 51
in pregnancy 165
lifestyle 159
promotion 205
risk of diabetes 51
extinction 46–8

family size 116
famine effect on birth size 165
farming *see* agriculture
fat:
at birth 84–5
deposition as fuel reserve 119, 160
in food 119
feminism 116
fertility 76–7
fetal growth:
constraint of 69, 132
dependence on nutrient supply 132, 165–6
effect on later development 147–8
regulation 164–6
fetal life:
effect of iodine deficiency 3, 8
exposure to sex hormones 144, 189
formation of uterus 69
formation of brain cells 129, 185
formation of eggs 182
formation of muscle fibres 35
formation of nephrons 128
gene silencing 62
haemoglobin 63
loss of plasticity 82
predictive systems 169
feudal system 109
finches 17, 38, 97–8
fire 45, 70, 99, 101
fitness *see* Darwinian fitness
Floreana Island 98
Flores Island:
elephants 64
hominids 100
Fogel, Robert 178
folic acid 206
food:
enrichment 119
high energy content 173
food technology 122
forecasting 89–90
formula feeding *see* bottle feeding
fossils 46–8
France:
Albi 158
Lamarck 54
language 71
life expectancy 178
newts 74
revolution 112
study of growth 152

Gairdner, Sir William 112
Galapagos Islands 17, 38, 97
Gambia 48, 73
gender:
differences in body size 55
differences in brain function 145
identity 145
roles 138
gene expression:
as factor in fetal calcium deposition 30
control by epigenetic processes 35, 62, 67, 68, 78, 163, 166, 169, 203, 206
during early pregnancy 166
influence of environmental signals 66, 202
pattern in twins 21
role in developmental plasticity 35
variation during development 63, 78
gene silencing 61–3
gene therapy 202, 206
gene-environment interactions 28–31
generalist species 39–41, 126
genes 19–22
genetic variation:
as driver of selection 53
as mix of genes 11
constraint on 24
diabetes 162

in breeding 59
in regulatory genes 125
local 4
necessity for evolution 22
persistence 24
protection by developmental plasticity 169
resistance to infection 23
sickle cell anaemia 127
genetics:
Modern Synthesis with Darwinian evolution 60, 66, 203
modern understanding 9, 32
role in eugenics 61
Genghis Khan 100
Genotype:
as basic template for development 198
definition 19–22
giraffes 54
global warming 122
Gluckman, Peter 2–6
glucose 160
goitre 2, 3, 5, 43
golden toad 200
grandmother
effect on child survival 73, 120
hypothesis 190–1
Grant, Peter and Rosemary 17
Gregg, Norman 164
growth:
human pattern 83–6
nutrient requirements 160
Gulf of Carpenteria 137
guppies 86–7

Hanson, Mark 95–96
Hawkes, Kristen 190
Hazda tribe 142
health care:
improved access to 112–13
reallocation of priorities 209–10
resources 192, 193
heart disease 163, 173, 176, 182, 195, 200, 202
height:
environmental factors determining 113
genetic factors determining 164
hibernation 161
Himalayas 1, 2, 212
Hippocrates 163
Hmong people 114
Holocaust 81
Homo erectus:
evolution 29, 100
maturation 84
migration 47, 100
use of fire 99
Homo floresiensis 100
Homo neanderthalensis 45, 47, 72, 100
Homo sapiens:
art 158
as generalist species 45, 47, 72, 100
conflict 158
effect on environment 108
evolution 29, 50, 94, 99, 100
life-history strategy 86
maturation 143
migration 45, 47, 100
origin 99
pubertal growth spurt 84
honey bees 35–6
hormone replacement therapy 191
Hubel, David 129
Human Genome Project 61, 201
hunter-gatherers:
diet 119
dispersal 101
energy use 120
feast and famine 107
food supply 103, 107, 152, 162
group size 85
infanticide 116
menopause 188
modern 102, 106, 110, 111, 149
myopia 197
skills 142–3
social structure 120, 137
thrifty genes 162

timing of puberty 149
transition to agriculture 117, 126
see also Palaeolithic humans
Huntington's disease 59
hypertension 173, 174

Ice Ages 45, 100, 104
immediate adaptive responses 82–3
imprinting 62
India:
childhood obesity 120, 159, 161
diabetes 119
hypertension 174
life expectancy 178
nutritional transition 119, 171, 174, 207
prevalence of metabolic syndrome 174
small babies 174
stunting 176
Industrial Revolution 120–1
infancy:
constraint of growth 133, 172
dependence on mother 84, 172
length 140
rapid growth in 84, 85
social behaviour 141
infant nutrition 172
infanticide 8, 27, 57, 116, 137, 149
influenza virus 23, 106
inheritance:
as tenet of Darwinian evolution 53
cultural 69–73, 207
epigenetic 15, 51, 52, 67, 125, 131, 198, 206
genetic 50, 51, 61
mechanisms 49–51
role in determining constitution of the individual 200
study by Mendel 60
insulin 171
Intelligent Design 7
intensification 104
intersex syndrome 57–8
Inuit 8, 45, 98, 197
Investment:
financial 94, 179, 193, 209, 210
life-history strategy 57, 86, 118, 130, 182, 185
parental 28, 140, 183
reproductive 139, 180
societal 176, 210, 211
iodine 3, 160
iodine deficiency:
adaptation to 200
fetal consequences 166
leading to cretinism 3, 4, 8
Sherpa 2, 4, 6
treatment 4
irish elk 56
irreversibility:
iodine deficiency and brain development 4
mismatch of timing of puberty 155, 157
of developmental plasticity 63, 128, 129
of gonadal development 82

Japan 170, 178
Johnson, Samuel 111

kairomones 10
kangaroo rat 41
Kanuri tribespeople 94–6, 123, 194
Koch, Robert 201
Koshima Islet 69–70
!Kung 102, 188

lactase deficiency 12, 105
lactose intolerance 12, 105
Lake Chad 94–6
Lamarck, Chevalier de 54
language:
as communication 69, 70
development 101
evolution 70–2
learning 129
role in cultural evolution 70–2, 76
sign 71, 137

use by Neanderthals 72
see also speech
Lascaux caves 158
learning:
as a form of developmental plasticity 129
during maturation 141–2
life expectancy:
decrease by metabolic syndrome 173
human 114, 115
in captive animals 189
in cities 110
of pilot whale 188
Palaeolithic 142, 149, 178, 183
secular trends in human 178
life-course approach to disease 204, 209, 210
life-course strategy *see* life-history strategy
life-history strategy 25–8
and timing of menopause 190
and understanding of disease causation 204
as adaptive response to environmental signals 83, 130
choice of 87
classes of 90–3
definition 25
direction by key genes 204
effect of correct forecast 89
effect of environment 90–3
human 28, 180, 183, 190, 199
trade-offs in 184
tuning by early environment 90, 93, 128, 199
lifespan:
maximum human 178
of non-human species 184
Palaeolithic 183
reasons for increase in human 181
Linnaeus, Carolus 67
Lisbon earthquake 111
liver 187
locusts 24
longevity:
and timing of menopause 184, 189
effect of reproductive success 91–2
genetic factors influencing 184
implications for social structure 192
of different species 85–6
Lysenko, Trofim 64–5

macaque monkeys 69–70
macronutrients 160
malaria 61, 115, 127, 165, 181, 194, 207
malocclusion 34–5
Maori:
arrival in New Zealand 101
effect of colonization 101
language 71
maple syrup urine disease 30
marsupial mouse 140
marsupials 92–3
Masai 137
mate choice 18
maternal behaviour:
effect on offspring 168
modulation of environmental signals 131
maternal constraint:
limitation of fetal growth 132, 170
reproductive competence 146
role in developmental mismatch 133
maternal diabetes 166
maternal maturity 142
maternal nutrition:
as fetal stressor 168
at conception 166
effect on birth size 165
effect on obesity and diabetes in offspring 174
maternal workload 165
mating strategies 26–8
maturation:
biological 138–43
cognitive 150
human 140–1
mismatch 143
psychosexual 144–6

psychosocial 138–43
McBride, William 164
media and marketing 155
melatonin 90, 144
menarche:
definition 143
in Palaeolithic 149
relation to physical growth 146, 176
secular trend in age 147, 153–4
see also puberty
Mendel, Gregor 60
menopause 188–91
adaptive advantage 190
as a consequence of increased longevity 189
as mismatch 188
definition 182
grandmother hypothesis 190
implications for hormone replacement therapy 191
in non-human species 188
mechanism 189
mercury poisoning 8
Meriam people 143
metabolic mismatch 173–5
consequences 173, 195, 200
developmental induction 171
nutritional transition 207
possibility of reducing 205
metabolic syndrome:
as consequence of metabolic mismatch 173
failure to be selected against 177
in developing societies 174, 175
possibility of elimination by natural selection 208
metabolism:
control by thyroxine 2–3
development 169
evolution in humans 126, 162
factor in longevity 184
kangaroo rat 41
locust 24
maternal, as determinant of fetal nutrition 80, 131, 165, 167
mismatch 173–5
of stress hormones 81
regulation 148
metamorphosis 74, 129
micronutrients 160, 166, 175
microvariation 20, 23, 29, 59
migration:
effect on growth and health 113
effect on nutritional environment 171
human 44–6, 113, 171
of birds 161
out of Africa 29, 50, 100–1, 122
Minamata Bay 8
mismatch:
approaches to reducing 205–8
as a new aetiology of disease 201
as a result of increased longevity 181–5
between brain size and lifespan 186
caused by bottle feeding 195
caused by changing demography 193
caused by urbanization and education 196–7
challenges 205–8
costs 195
detoxification 187
developmental 130, 133, 202
genetic factors influencing 177
imposed on other species 199
maturational 151, 154–7
menopause 189–90, 195
metabolic 173–5
need for further research 210
nutritional 174
of humans in modern world 122–3
of physical and psychosocial maturation 151, 154–7
of social structure caused by increased longevity 193
psychological 197–8
timing of puberty 151, 154–7
mismatch paradigm 7, 11–13, 194, 199–200, 204, 211
mitochondria 100, 160
mitochondrial DNA 100

Modern Synthesis 60, 66, 203
monotocous pregnancy 26
monotremes 93
Monteverde Cloud Forest 200
mortality:
  infant 48, 91, 110, 181, 207
  neonatal 137, 172
mouse:
  agouti 67–8
  laboratory 52, 189
  lifespan 86, 189
  marsupial 140
  wood 86
mutation:
  as copying error in DNA 20
  Boroola 19
  causing sickle cell trait 127
  driver of variation 125
  in sperm 51
  lactase deficiency 12
  neutral 18, 125
  of influenza virus 23
  selection of favourable 34
  types 20
  use in population tracking 100
myopia 197

natural medicines 114
natural selection:
  action 17
  adaptation to environment 10, 33, 53, 87
  and ageing 86
  as part of Modern Synthesis 60
  as solution to mismatch 208
  determining pelvic dimensions 146
  elimination of mutations 29
  giraffe 54
  in recent human genetic changes 126
  relation to artificial selection 60
  relation to sexual selection 55
  role in cultural inheritance 70
  time frame 39
  versus effects of environment 64
Nauru Island 8
Neanderthals *see Homo neanderthalensis*
Neel, James V. 162
Neolithic 102, 113, 121
Nepal 1
nephrons 128
Nesse, Randolph 203
neurodegenerative diseases 182, 186, 202
New Zealand:
  bird migration 161
  settlement by humans 101
newts 74, 88
niche *see* environmental niche
Nigeria 94–6
North America:
  kangaroo rat 41
  meadow vole 10
  mountain vole 88
  settlement by humans 101
  see also USA
nutritional environment:
  at conception 166, 176
  during pregnancy 80, 174
  fetal 131–3, 170–1
  human 98, 113, 118, 122, 171, 195
  infant 84, 89, 148
  maternal, as determinant of fetal nutrition 167
  response to poor 91
  scurvy 12
nutritional transition 51, 118–20, 174, 175

obesity 173, 158–177
  bottle feeding 196
  China 161
  component of metabolic syndrome 173
  consequence of metabolic mismatch 135, 173, 195
  environmental factors influencing 27
  genetic factors influencing 27
  in sub-Saharan Africa 119
  increase in prevalence 159, 167
  India 120, 207

link with reduced energy expenditure 161
parental behaviour as cause 51
reversal in animal model 176, 206
social effects 159
thrifty genotype 162
see also childhood obesity
osteoporosis:
as consequence of adverse prenatal environment 185, 210
as degenerative disease 182
effect of vitamin D receptor gene polymorphism 202
societal costs 196
vitamin D deficiency 44
ovary:
development 82
egg formation 50, 77, 124, 144, 146, 190
menopause 188–90
puberty 145
oxidative stress 86

Pakistan 210
Palaeolithic humans 101–3
agriculture 105
clan size 143, 197
diet 103, 118
environment of evolutionary adaptedness 103
food supply 149
language 117
life expectancy 149, 178, 183, 191
life-history strategy 183
menopause 188
plant remedies 114
puberty 148
sexual selection 57
social structure 103, 120, 182
weaning 133
see also hunter-gatherers
panda 39, 41
pandemic:
diabetes 174
influenza 23, 106, 114, 127
Parkinson's disease 186
Pasteur, Louis 201
pelvic canal:
relation to timing of puberty 146, 176
shape change by bipedality 132
size in relation to birth 85, 92, 93, 132, 140, 164
Pennsylvanian meadow vole 10
Phenotype:
definition 19, 21
match to environment 36, 38–9, 83, 127, 133
plasticity during development 75, 82, 90, 198, 203
selection 22, 31, 32, 42
variation 23–5, 59, 63
pituitary gland:
control of thyroid hormone secretion 3
puberty 145, 146
placenta:
as transducer of environmental signals 80, 89, 90, 131, 164, 169
calcium transport 30
dysfunction 90, 131, 165, 167, 170
epigenetic markers 206
hormones 167
transport of nutrients and oxygen 63, 131, 165, 167, 168, 170
plants:
as food source 24, 36, 46, 97
breeding 23, 59, 65
climate change 44, 46
defences 187
domestication and agriculture 60, 64, 104, 105
genetics 67
remedies 114
vernalization 65
polyandry 5, 27, 72
polymorphism 20, 21, 59
polyphenism 35
polytocous pregnancy 26
population density 152

precocious puberty 154
predation:
as component of environment 40, 46, 80, 87, 90, 91, 128
as reason for extinction 56
defence against 6, 18, 25, 36, 38, 42, 70, 187
effect on life-history strategy 10, 18, 81
prediction:
as component of developmental plasticity 168, 180, 184
changing to address mismatch 175
defined as biological expectation 9
importance of fidelity 10, 89, 90, 131, 171
predictive adaptive responses:
definition 82, 167
fidelity 131, 169, 198
role in developmental plasticity ad adaptation 89, 168, 171, 172
role in mismatch 83, 171, 184
signals 89
pre-eclampsia 170
pregnancy:
birth defects 58, 59, 164
diabetes 79, 167, 174
dieting in 170
environment during 80, 90, 144
exercise in 165
monotocous 26, 91
nutrition during 48, 68, 80, 147, 165–70, 174–5, 206
optimal timing of first 176
optimizing outcome 166, 170, 176, 206–10
polytocous 26, 92, 140
spacing 85, 88, 116, 147, 183
stress during 68, 81
teenage 207
promoter region 22
pubertal growth spurt:
first documentation 153
gender-specific timing 146
uniqueness in humans 84, 116
puberty:
advancement by poor fetal nutrition 147
delay by poor childhood nutrition 147
effect of endocrine disruptors 148
mismatch in timing 151, 154–7
Palaeolithic 148–9
pathological 154
timing in relation to nutrition 147–9
see also menarche
puberty rites 137–8
public health:
effect on growth 153
effect on lifespan 114, 194
improvement 112, 153
measures to optimize pregnancy 166, 175, 208

rabbits 38–9
rats 6, 40, 140, 176, 206
rebound in fatness 172
recessive inheritance 58
religion:
as part of cultural evolution 72–3, 101, 109, 111, 112, 114, 143
as source of conflict 158
cultural inheritance of 52, 114
reproduction:
as component of life-history strategy 86, 91, 118, 177, 180, 182, 184, 185, 190
assisted 77, 117, 127, 164
control over 115, 116–17, 152
cultural inheritance and 72, 156
in monotremes 93
mammalian 140
menopause 183, 189, 191
seasonal versus continuous 144
span in Palaeolithic humans 149
reproductive success:
advantage of match 83
effect on longevity 91–2
respiratory infections 115
retirement 192

reversibility 176
rickets 43, 163
Rome 107–8
rubella 79, 164
Russia:
  scientific thinking 24, 32, 64, 65, 203
  silver fox 25

salmon 25, 86, 126, 139
San Andreas Fault 79
Sapolsky, Robert 137
Schmalhausen, Ivan Ivanovich 32, 65, 203
scurvy 12
secular trend:
  age of menarche 153
  height 153
selection:
  against ageing 180, 182, 190, 191
  artificial, *see* artificial selection
  as tenet of Darwinian evolution 53
  basis 22
  environment as driver 63
  natural, *see* natural selection
  of ability to adapt 32, 33, 38, 39, 125, 198
  of human ancestors 103, 118
  of life-history strategy 25
  of modern humans 126, 127
  role in speciation 46
  sexual, *see* sexual selection
  substrate for 31, 38, 208
  time frame for 39
semelparity 139
September 11, 81
settlement 106–7
sex determination 57–8
sex education 156–7
sexual behaviour 145
sexual orientation 145
sexual reproduction 18
sexual selection:
  Darwin's writing on 55
  giraffes 55
  Irish elk 56
  of human characteristics 57, 146
  peacock 56
shamanism 113
Sherpa:
  adaptation to environment 33, 41, 43
  as example of match 212
  as example of mismatch 48, 134, 194, 200
  effects of iodine deficiency 2–8, 166
  social structure 1, 2, 5
Siberia 24, 65, 104
sickle cell anaemia 127
sign language 71
silver fox:
  artificial selection for tameness 25
  domestication 104–5
skeletal muscle 160, 169
skeleton:
  growth at puberty 84, 146
  of hunter-gatherers 84, 104, 149
  rickets 43
smoking:
  during pregnancy 176
  effect on nutrient transport across placenta 90, 170
  health warning 201
social behaviour:
  adolescence 141
  maturation 141
social structure:
  as component of life-history strategy 25, 40
  effect of agriculture 151–2
  effect of mismatch 13
  evolution of human 106–13
  honey bee 35
  human 45, 72, 93, 107, 109, 122, 125, 198
  implications of increased longevity 192
  in cities 107, 121
  modern 111, 156, 192
  Palaeolithic 143, 182, 183

primate 57, 142
Sherpa 5
society:
complexity 107, 135, 151, 154
demography 192
Kanuri 96
maturational mismatch in modern 154–7, 195
Nauru 8
resource allocation 195, 196, 204, 205, 207, 211
Sherpa 1, 4
skills for 106, 152
stratification 152, 153
sociobiology 28, 198
socioeconomic status 173
South America:
cinchona tree 114
settlement by humans 101
spadefoot toads 36–7
specialist species 39–41, 97, 126
speciation 24, 29
speech:
and language development 71
brain pathways controlling 70
see also language
sperm:
epigenetic changes 166
formation 76
start of life 75–7
stem cells:
brain 185
controversy 77
embryonic 63, 66, 67
stress:
competition or predation 37, 81
during pregnancy 68, 80, 81, 164, 168, 169
employment 121
jaw and malocclusion 35
oxidative 86
timing of puberty 148
stunting:
as maladaptation to undernutrition 175
India 174, 176
maternal 174, 176, 208
sub-Saharan Africa:
bilharzia 95
metabolic syndrome 175
nutritional transition 119, 159, 175
sickle cell anaemia and malaria 127

tadpoles 129
Taiwan 197
Tasmania 5
technology:
aboriginal 102
agricultural 73, 105, 117
energy-saving 120, 122
epigenetic 208
for tracking population movement 100
manipulation of environment 98, 104, 108, 109, 122, 208
modern 75, 98, 112, 117, 121, 126, 205, 208
Palaeolithic 45, 100, 101, 117, 121
reproductive 77, 117, 127, 156, 164
termites 45, 96
testes:
chimpanzee 57
development 55
fetal 58, 82
parasite 26
puberty 145
sheep 144
sperm formation 76
testosterone 144
testosterone:
puberty 145
sex determination 55, 58, 144
thalidomide 79, 164
thrifty genes 162, 175
thyroid hormone *see* thyroxine
thyroxine 2, 3, 4, 43, 78, 160
Tibet 1
tiger snake 33–4

timing:
  cognitive maturation 142
  evolutionary change 46
  life course 85
  menopause 188
  pregnancy 116
  puberty 138, 141, 142, 147–51, 153, 155, 195
  rebound of fatness 172
tissue differentiation 77–8
toadflax 67
toads:
  developmental plasticity 36, 37
  golden 200
  spadefoot 36, 37
tools:
  chimpanzee 70, 141
  developmental plasticity in evolution 79, 128, 177
  epidemiological 210
  epigenetic 206
  hominid 45, 47, 70, 99, 101, 102
  Inuit 98
  molecular biological 77, 100
trade-offs:
  and ageing 185, 186
  as elements of life-history strategies 91, 182
  effect on brain development 185
  in human life-history strategy 184
  menopause 190
  see also costs
traditional medicine 114
trait:
  ageing 177
  definition 10
  genetic influence 53, 60
  male 55
  selection 17, 25, 38, 198
  sickle cell 127
transcription factors 22, 53
transgenerational inheritance 66–9
transitions:
  adolescence 138
  in human history 117
  nutritional 118–20
  social 120–2
  workload 118–20
Tuareg 8
tuberculosis 114, 115, 201
type 2 diabetes *see* diabetes

UK:
  childhood obesity 159
  goitre 6
  health 113, 115
  life expectancy 114, 178
  maternal age at birth 116
  social reform 153
  spelling 20
  tuberculosis 114, 115
  see also *England*
universal suffrage 116
USA:
  American Revolution 111
  bodyweight 159
  life expectancy 178
  migration 113
  nutritional conditions 113, 178
  shamanism 113
  spelling 20, 172
  see also North America

vaccination 114, 115, 194, 201
variation 22–5
  as tenet of Darwinian evolution 53
  genetic, *see* genetic variation
Venus figurines 158
vernalization 65
viral infection 114
visceral fat 160
vitamin C 12
vitamin D 30, 43, 44, 163, 202
voles:
  mountain 88, 89
  Pennsylvanian meadow 89
Voltaire 111

Waddington, Conrad 32, 65, 203
wallaby 88
Wallace, Alfred Russell 21

Waterlow, John 175
Weaning:
  human dependency beyond 133, 140, 172
  in mammals 140, 196
  timing in humans 84, 85
Weisel, Torsten 129
welfare systems 112–13
Wiedemann-Beckwith syndrome 62
Williams, George 203
Wilms tumour 62
Workload:
  in pregnancy 165, 175, 176
  transition 118–20
  variation with disease 96

X chromosome 20

Y chromosome:
  as genetic marker in population studies 100
  inheritance 100
  sex determination 20
  testis formation 55